Praise for **INDUSTRI**

T0288862

"*Industrial DevOps* unveils the hidden strategies behind the remarkable success of the world's top-performing companies. Dr. Suzette Johnson and Robin Yeman have masterfully crafted a comprehensive and actionable guide, empowering organizations to establish teams and capabilities that surpass the competition in both insight and agility."
—Kyle Fox, SOSi CTO

"Applying established DevOps processes and practices to cyber-physical systems can present significant challenges or limited practices. I am thrilled to see that *Industrial DevOps* not only recognizes these challenges but also provides actionable guidance based on well-practiced approaches to overcome them."
—Hasan Yasar, Technical Director and Teaching Professor, Carnegie Mellon University

"Suzette and Robin have created an essential guide to building cyber-physical systems. . . . This book is a great addition for anyone wanting to deliver higher-quality products faster and more predictably. Complete with thought-provoking discussion questions and coaching tips, this book is a significant work that kicks off the era of Industrial DevOps!"
—Jeff Boleng, Aircraft Cybersecurity Lead, Joby Aviation

"*Industrial DevOps* is an important book and topic for anyone helping steer their cyber-physical organization into the age of digital engineering. Using in-depth case studies and sweat-equity observations from years of relevant product development, this duo of top-level engineers helps your organization become both agile and speedy."
—Dr. Steve Simske, Systems and Mechanical Engineering, Colorado State University

"*Industrial DevOps* clearly articulates what businesses must do to revolutionize the way they work and stay relevant in the digital age and beyond. The book is brilliantly laid out, explaining why change is necessary, identifying key principles and concepts to consider when implementing change, and pulling it all together using real-world examples. It is a must-read for all Lean-Agile professionals."
—Gabriela Coe, Northrop Grumman Fellow

"Crashing through the myth that 'agile is only for software,' *Industrial DevOps* distills decades of Agile practices into a functional model, arranging real-world examples into forward-facing guidance for the development of cyber-physical systems."
—Braxton Cook, Aerospace Professional

"Agile practices evolve fastest when they are based on a stable and timeless foundation. Suzette Johnson and Robin Yeman present the foundation needed for organizations to adopt—and evolve—Industrial DevOps in a way that helps them realize greater business agility."

—**Luke Hohmann**, Co-Author *Software Profit Streams*™, SAFe Fellow

"Dr. Suzette Johnson and Robin Yeman have pushed the Agile and DevOps envelope with this wonderful book. *Industrial DevOps* is a compelling guide for those interested in DevOps at scale—not only applied to cyber-physical systems but to any large, complex ecosystem. If you're looking for a clear handbook with practical advice backed up by time-honored principles, this is a must-read."

—**Sanjiv Augustine**, Founder and CEO, LitheSpeed
and the Agile Leadership Academy

"If you are leading or managing complex projects or even just a working member of a complex product team, this book is definitely worth a read. You are certain to gain insights and ideas that will push your Agile up another notch."

—**Nathan G. Christensen**, Senior Manager, Engineering and Scientific Methods,
Northrop Grumman Space Systems

"Any executive, manager, technical leader, or engineer who does not want to be left behind should read this book and study its methods. *Industrial DevOps* does an excellent job of combining the *why* with the *how* in a practical and understandable way."

—**Reggie Cole**, Lockheed Martin Senior Fellow

"Using a realistic scenario to provide specific examples of best practices, *Industrial DevOps* is the only book on the market that addresses how the full Agile DevOps life cycle applies to the aerospace industry. *Industrial DevOps* is a must-read for managers, technical leads, and developers who work in this space."

—**Gordon Kranz**, President, Enlightened IPM

"This book can serve as a textbook, complement existing training and coaching programs, or simply be a good recommended read to accelerate your journey toward business agility and digital transformation."

—**Sandra J. Forney**, DEng, Professional Lecturer and
Leader in Digital Transformation

build better systems faster

INDUSTRIAL DEVOPS

Dr. Suzette Johnson
and Robin Yeman

Forewords by **Mik Kersten** and **Dean Leffingwell**

IT Revolution
Independent Publisher Since 2013
Portland, Oregon

25 NW 23rd Pl, Suite 6314
Portland, OR 97210

First Edition
Printed in the United States of America
28 27 26 25 24 23   1 2 3 4 5 6 7 8 9 10

Cover design by Devon Smith Creative
Book design by Devon Smith Creative

Library of Congress Control Number: 2023027080

ISBN: 9781950508792
eBook ISBN: 9781950508808
Web PDF ISBN: 9781950508815
Audio: 9781950508822

For information about special discounts for bulk purchases or for information on booking authors for an event, please visit our website at www.ITRevolution.com.

INDUSTRIAL DEVOPS

CONTENTS

FIGURES & TABLES

Chapter 6

Chapter 7

Chapter 8

Appendix C

Foreword *by Mik Kersten*

With the acceleration of software and the advent of generative AI, we are witnessing the fastest pace of change in our careers. This puts tremendous pressure on both organizations and individuals to learn and adapt in a rapidly shifting technology landscape.

Since the 2000s, we have seen digital natives create the delivery and feedback loops that have enabled them to capture a massive portion of the global market. Since then, much has been written about digital transformations, and an abundance of guidance provided for how large enterprises should go about becoming software innovators. However, comparatively little has been written about how organizations building physical systems should transform. That's where Industrial DevOps comes in.

A small number of organizations that take a software-centric approach to building physical systems have won a disproportionate part of the market for physical products. Apple and Tesla alone have disrupted countless slower moving incumbents and amassed a dizzying amount of market share in the process. What we need for a thriving economic landscape is for the rest of the world's organizations to match this pace of product delivery and innovation.

What *Industrial DevOps* does so eloquently is make the case for why this shift is not only possible but critical for your organization to thrive. In addition to providing concrete guidance, this book has a wealth of case studies that will show you what good looks like regardless of your industry or scale. The Joby Aviation case study demonstrates how these principles apply to newer and nimbler companies wanting to move fast against entrenched competitors. At the other end of the spectrum, the Lockheed Martin U-2 Dragon Lady case study demonstrates how the same principles can be applied to bring machine learning and DevOps via Kubernetes containers to a reconnaissance aircraft created back in the 1950s. *Industrial DevOps* will provide you and your organization with numerous examples that ground the principles presented in guidance that you can start actioning today.

There is no one I know better suited to put forth these best practices and methodologies than Robin and Suzette. Their close involvement with the IT Revolution DevOps Enterprise community for the past decade has given them a perspective that few others have. More importantly, thanks to their groundbreaking work applying the principles of Agile and DevOps to cyber-physical systems, they have continually expanded our communities' understanding of how to bring these benefits across horizons that span physical systems and the longer time horizons necessitated by hardware.

Suzette and Robin have collected that wisdom in this book, and they demonstrate how the principles of DevOps can not only move mountains within your organizations but can go as far as moving interstellar objects. Quite literally, as one of the case studies in the book is NASA's DART mission, which applied continuous delivery (CD) automation to hardware-in-the-loop (HIL) and software-in-the-loop (SIL) systems to successfully change the trajectory of an asteroid. This was the first time that humanity had changed the motion of an interstellar object. If changing the direction of your organization seems like a harder task than changing the direction of interstellar objects, Robin and Suzette provide an invaluable summary on how to surmount organizational objections and make the economic case to leadership.

I have had the pleasure of working with manufacturing organizations that have participated in the DevOps Enterprise community early and were some of the first adopters of the principles outlined in this book. The journey of one of those organizations, BMW Group, is chronicled in *Project to Product*, where I recount my visit to the Plant Leipzig.

I happened to be reading the manuscript for *Industrial DevOps* while making a visit to BMW Group's Munich plant, hosted by Ralf Waltram (VP IT Delivery DevOps for Production at BMW Group). Ralf helped me come up with the title of *Project to Product* as he was leading the BMW Group's shift to Agile. He made this anniversary tour even more intimate, showing me an amazing combination of digital workstations combined with active car assembly via human and robot collaboration.

I could not believe the progress that I saw since visiting the Plant Leipzig exactly six years earlier. The manufacturing plant was producing electric, hybrid, and combustion engines all on a single line, just in sequence (JIS) of orders. I had not fathomed that could be possible given the extremely different physical components and calibrations required for one body that housed a massive battery being followed by another that contained a combustion engine.

As Ralf pointed out, managing this kind of real-time complexity would simply not be possible without a software-centric approach based on digital twins, simulation, and platform architectures. It will be exactly these and other leading-edge practices of product development innovators that you will learn about in this book.

Each technological revolution builds on the pioneering work of the organizations that paved the previous one. We have entered the Age of AI, which I believe will provide productivity gains of at least an order of magnitude for software development and architecture. Organizations that take a software-centric approach to their product development will be able to quickly reap those massive productivity gains and to deliver value to their customers or citizens at an unprecedented rate. Those restricted by the constraints of waterfall development will not have the infrastructure for innovation needed to leverage the benefits of these tectonic shifts.

Industrial DevOps is your guide for the changes that you need to put in place to reap the benefits of digitization and AI. The guidance in the book ranges from architecting around flow and value streams to the simulation and automation needed to 10x your hardware and software feedback loops. I firmly believe that the organizations adopting the practices in this book will outperform and outlast those that do not lean into the change and learn to move at the speed of DevOps. I hope you use this book to build the road map for your team's journey. Whether that leads you to the stars or to your organization's north star, you are about to embark on it with two of the best and most proven guides in the industry: Robin and Suzette.

Enjoy!

—Mik Kersten
CTO of Planview and author of *Project to Product*

Foreword *by Dean Leffingwell*

Sometimes major breakthroughs in technology development happen quietly. They might not initially make headlines, but their impact on how we innovate is indelible. Agile development is a great example.

It's hard to believe it's been over twenty years since Agile, and later DevOps, emerged as a better way of building software and systems. Thankfully, I can now say that I've spent more time working in Agile than in my prior years applying waterfall. But I must admit, it took me a couple of years to join the Agile party.

My earlier developer and executive years were spent primarily developing high-assurance systems such as medical devices, DNA sequencers, heavy-duty industrial robotics, patient monitoring systems, aerospace systems, and the like. The development of those systems gave me great respect for documented requirements and design, traceability, meeting imposed regulations, and all the other aspects of delivering systems that did exactly what they were supposed to do and no more.

Agile values and principles didn't address those issues; initially, I wasn't sure they would be useful for that kind of work. But I then had the opportunity to coach several pure software startups. Free of all those high assurance constraints, I learned and coached Agile with several new companies. One was Rally Software, where I was mentored and influenced by Ryan Martens, an early and influential Agile proponent.

Simply put, I was blown away by the power of Agile. It was unlike the methods that came before. It was a quantum leap forward. What's more, once I experienced being on an Agile team, I vowed never again to work in a situation where I couldn't be on one. That led me to a new purpose: bringing the power of Agile to those building the world's largest and most important systems. This new passion led me to write *Scaling Software Agility: Best Practices for Large Enterprises* and *Agile Software Requirements: Lean Requirements Practices for Teams, Programs, and the Enterprise.*

Even then, the application of Agile—and scaled Agile—was primarily constrained to software systems. The challenge of building high-assurance and cyber-physical systems with Agile never left my mind. I continued to research and observe how scaled Agile methods could be applied to these systems. And when it became apparent that Agile development methods produced natively and fundamentally higher quality outcomes, I reached another tipping point.

Working with Dr. Harry Koehnemann and other industry experts, we branched the Scaled Agile Framework® (SAFe®) into a beta version of SAFe for Lean Systems Engineering. But as early testers soon discovered, it became evident that the principles and practices of building these systems were strikingly similar to large-scale software development. With that knowledge, we instead folded those learnings into the new SAFe Enterprise Solution Development competency.

Two of the industry experts who helped guide us through that exploration journey are the authors of this book, Robin Yeman and Dr. Suzette Johnson. I found their perspectives and experiences to be invaluable to my new purpose. In their respective roles at Lockheed Martin and Northrop Grumman, they had been working—both cooperatively and competitively—to extend Agile, DevOps, Lean, and SAFe to build everything from satellites to submarines, as Robin likes to say.

I have always been impressed by their experience, insights, grace, and gumption. You can imagine it was no easy journey. They worked for very large, very conservative, and very successful companies, who, in turn, delivered systems to the world's largest bureaucracy, the US government. Undoubtedly, they found significant impediments blocking their progress every day. And they were not just working the "easier" pure software stuff; they were applying these innovative and advanced methods to very large-scale and safety-critical cyber-physical systems. These systems contained combinations of software, mechanical, firmware, electrical, electronic, hydraulic, propulsion, optical, sensors, actuators, and physical devices. Moreover, these systems interacted with other equally complex systems in extensive communication networks. Tough stuff, to put it mildly.

Meanwhile, the DevOps movement was gaining momentum. I vividly remember the fourth annual DevOps Enterprise Forum in Portland, Oregon, in 2018. This private event was sponsored by Gene Kim and IT Revolution. Invited experts met and collaborated in what Gene described as a "scenius" (communal genius). Teams collaborated, wrote, and published (mantra: "in DevOps, we ship") guidance papers on the most challenging and relevant topics to the industry.

I joined a working group instigated by Robin and Suzette that was keen to work on this topic. Initially, the task was daunting. How could anyone describe the application of Lean, Agile, DevOps, and Systems Thinking bodies of knowledge to building systems that were so diverse and complex that the systems themselves were hard to describe in any detail, much less document and validate with any rigor? The conundrums were many: continuous integration meets the laws of physics; fast iterations meet the laws of overloaded machine shops; fuzzy user stories meet the laws of documented software requirements specifications.

But a breakthrough came when the working group agreed that while the problems and experiences were vastly different, and comprehensive guidance might be unattainable, there was another way to move the industry forward via the application of common principles. This W. Edwards Deming quote inspired us: "[The problems] are different, to be sure, but the principles that will help to improve the quality of product and service are universal in nature." That's when we knew we could create something worthwhile. We'd take a stand on agreeing to a set of principles—underlying truths that could reliably inform the methods and practices needed to build these diverse and complicated systems.

That effort resulted in the first published paper on the topic entitled *Industrial DevOps: Applying DevOps and Continuous Delivery to Significant Cyber-Physical Systems*. Further papers ensued, and Robin and Suzette wrote this book as a logical and much-needed outcome. Part II of this book highlights the evolved principles which form the core of Industrial DevOps.

But perhaps one of the most valuable aspects of this book is that it is not just about abstract principles. The ideas described here are augmented by real-world and exemplary case studies, practical tips, and guidance. I can say for certain that applying Lean, Agile, DevOps, and Systems Thinking to the creation of the world's most important systems is now more attainable. Doubters and naysayers may well abound in industry and government, and perhaps some will never give up their more prescriptive and stage-gated waterfall habits. But those with an open mind can now see the possibilities of building these systems faster, with higher quality and lower cost, and, importantly, enhancing the joy in doing.

And what could be more important than applying these principles to the very systems that provide society with security and safety? Surely that would truly make the world a better and safer place. We will soon owe some of that to Robin and Suzette.

This is an important work; as an industry and a society dependent on these future systems, we are lucky to have it.

—**Dean Leffingwell**
Creator of SAFe®

Preface

Suzette: My first experiences with Agile, specifically Scrum and eXtreme programming, started in 2005. The mission called for faster delivery of software capabilities. We had to find a way to manage the fast pace of change, and the practices uncovered from the Agile community seemed like the right fit for our environment. This was a large-scale implementation of many teams of Agile teams who were very successful at delivering at speed of mission. My Agile experiences had always been at scale.

My journey to Industrial DevOps started around 2015, when I was asked to help a software development team of about sixty people implement Agile and DevOps. The project was a special radar program. Its unique capabilities were designed to help the defense community provide special data and reporting for critical decision-making activities. However, these large radar systems require more than software. The software must run on hardware. And therein was the challenge.

Before I came on board, the project had been using a waterfall approach to development. The project was "in the red," and not only was it running behind schedule, but the slip in schedule was also getting a great deal of attention from customers and management. Concerns were mounting. The risk to try something new was low, and management's mindset was, "Well, we've tried everything else—we may as well try this Agile/DevOps thing." So, we did.

Over the course of a year or so, things turned around. The software team did an outstanding job of adopting the new practices, and the project (from a software perspective) was back on track. But, while the software was reporting green, the project as a whole was still reporting crimson red. How could this be? This is when I realized the misstep that had been made. Agile/DevOps principles and practices had not been applied to the whole value stream. We had neglected to bring the hardware team, and a couple of other teams further down the value stream, along with us as we adopted these new ways of working.

Around this time, I had also been reading the book *The Goal* by Eliyahu Goldratt. *The Goal* is a novel about a Lean manufacturing plant. It teaches concepts such as improving flow, productivity, and continuous improvement on the manufacturing line. The analogy that could be drawn between the learning from Goldratt and what I just experienced was serendipitous. It truly was a moment of epiphany.

Much of the effort I had spent with the team was focused on improving the flow of software development and removing bottlenecks at the *software* integration points. The software teams had shifted to an Agile and DevOps approach while the rest of the functions were still using waterfall practices. While the results from the software were good, the overall effort was not. And that is when I thought back to what I had learned from Goldratt: if you build in all your efficiencies before the bottleneck in the assembly line and you still can't deliver to your customer, then your efficiencies are likely an illusion. The bigger bottlenecks were *between* software and hardware.

The real goal was to deliver the product to the customer, not to improve software development. We had neglected to assess the whole picture. What was needed was a systems view of the problem and to understand how all the teams were organized around the flow of value. By understanding and visualizing the full value stream, we were able to prioritize improvement efforts and focus on the highest-prioritized needs, wherever they fell within the value stream.

In the summer of 2017, I began volunteering at an Agile conference in San Diego. I already had experience with scaling Agile in cyber-physical systems, and I wanted to learn more. I was curious if anyone else was having similar experiences. That led me to meet Joe Justice.

He had been very successful with Wikispeed, an automotive manufacturer of modular cars using Scrum and eXtreme programming practices as applied to hardware. He invited me to a Scrum for Hardware workshop he was giving at Bosch Engineering, just outside of Stuttgart, Germany, later that fall. It was an amazing experience. Bosch was implementing Scrum in their hardware development, and they were gracious enough to include me in their train-the-trainer course and share their approach and steps toward scaling across larger parts of the organization. My assumptions were validated in the art of the possible.

Robin: My experience with Industrial DevOps concepts has been an evolution over many years, beginning with the transition to Agile in 2002. Every system that we build and maintain in the aerospace community is large-scale, cyber-physical, and often safety critical. I found time and time again I would have teams doing an excellent job completing the software rapidly by using Agile practices, only to find that the software would be shelved, waiting for the hardware to be completed. By the time the hardware had been completed, we found that the software needed to be adapted based on the actual hardware and the changes in the mission environment. Bottom line: we were not looking at the whole value stream and addressing the theory of constraints. Any improvement we made in the process was not resulting in changes in the outcome for our customers. This problem was replicated in every environment I supported, from radar to ships to aircraft.

In 2013, I began researching a variety of materials on implementing Agile for hardware from several pioneers in the field, including Joe Justice with team Wikispeed and Gary Gruver with Hewlett Packard.

In 2014, I implemented our first Agile for Hardware pilot building a missile with an existing legacy program. We experienced many benefits using Agile ways of working, including reduced rework, shorter lead times, and increased transparency. At Lockheed Martin, we built an Agile for Hardware workshop and training materials, which allowed us to continue to experiment with Agile practices on hardware projects across multiple domains.

While not every experiment was successful, it became clear that we needed to expand Agile beyond hardware and software and to the entire value stream if we were going to maximize benefits to the end customer.

Over the course of the next few years I was able to identify key lessons learned, which included:

- **Agile is an empirical life cycle:** it needs to include all of product development to succeed.
- **Focus on the principles:** Agile/DevOps principles apply to all types of work.
- **Context matters:** Agile/DevOps utilizes many patterns; not all of them are applied to every area.
- **Begin with the problem** you are trying to solve, not a process.

Suzette and Robin: We soon had the opportunity to discuss what we were seeing at the fourth annual DevOps Enterprise Forum in Portland, Oregon, in the spring of 2018. At this private event, Gene Kim (best-selling author, researcher, and multiple-award-winning CTO) invites industry leaders and experts to come together for a couple of days to discuss the issues at the forefront of the DevOps Enterprise community and to put together guidance to help us overcome and move through those obstacles.

So, we went and made our pitch to the group:

The Big Idea: DevOps for large, complex systems utilizing both hardware and software.

Topic: Shifting large legacy organizations with ingrained traditional management and engineering practices into a DevOps flow of continuous delivery of value in large, complex systems with both hardware and software development.

Problem Statement: We recognize the benefits of DevOps in IT/software-centric environments. Can these same benefits be recognized in the development and engineering of large, complex systems that encompass both hardware and software development (like an autonomous car)?

Following our pitch, a team of leaders* from across the industry joined us to address this problem space. For the next several months, we worked on defining and publishing our first paper on the subject: *Industrial DevOps: Applying DevOps and Continuous Delivery to Significant Cyber-Physical Systems.*[1]

In the development of this paper, we agreed upon several things:

1. These software/hardware systems would be referred to as "cyber-physical systems."
2. We identified the success patterns we were seeing in the application of Lean, Agile, DevOps, and trends in digital engineering in software and hardware.
3. We pulled from existing bodies of knowledge, including Lean, Lean Startup, systems thinking, design thinking, systems engineering, model-based systems engineering, Agile, and DevOps.
4. We named the integration of cyber-physical systems and the success patterns "Industrial DevOps." *Industrial* pulled in the concept of hardware and manufacturing, while *DevOps* emphasized flow and fast feedback from software.

---

* The initial team included Diane LaFortune, Dean Leffingwell, Harry Koehnemann, Dr. Stephen Magill, Dr. Steve Mayner, Avigail Ofer, Anders Wallgren, and Robert Stroud.

Over five years, we researched and implemented our ideas, producing five papers as we refined the concept of Industrial DevOps. Finally, it was clear that it was time to collect our experiences and the refined concept into a book to help instruct others on how to build cyber-physical systems using Industrial DevOps principles. While this book is the culmination of years of experience, experimentation, failures, and successes, we recognize the journey is not over. As we broaden our experiences and learning, sometimes it feels like it is just the beginning. We expect the concepts presented here will continue to evolve over the coming years, and we firmly believe these principles will help businesses not just survive the digital age but thrive in it.

Introduction

It is no secret that we are living in the digital age. Technological advancements continue to grow at an unprecedented rate. Digital capabilities are impacting our businesses, making competition fiercer and increasing the need to expedite time to market. To succeed today, organizations must rapidly improve, innovate, and adapt to changing market demands or lose their foothold in the market.

A key imperative for companies to be successful is *speed to market*. And one of the clearest ways to achieve this is by updating our old ways of working (waterfall) with more Agile ways of working. In the digital world, countless organizations have achieved this speed through Agile and DevOps practices. But Agile and DevOps ways of working aren't just for software companies or digital natives. Today, these ways of working are crucial for all businesses, including those that create cyber-physical systems (combining software with hardware and firmware). And yet, few of these companies have successfully adopted these ways of working, putting the majority in the dangerous position of falling behind their competition.

The Need for Speed

In the 16th Annual *State of Agile Report*, which surveyed more than three thousand people from across industries and countries, 52% of respondents reported that accelerating time to market was a key benefit of transitioning to Agile ways of working. This benefit was coupled with organizations' desire to move quickly while still remaining predictable. In addition, 47% of respondents reported that Agile teams are measured by the speed of delivery, which highlights organizations' continued desire to deliver in shorter lead times.[1]

Over the past decade, Agile and DevOps have produced dramatic changes in the way software teams develop and deploy value to their customers. According to a 2019 *Harvard Business Review* report, organizations that have implemented Agile and DevOps practices demonstrated "increased speed to market (named by 70%), productivity (67%), customer relevance (67%), innovation (66%), and product/service quality (64%)."[2] Data from a 2020 report from the Standish Group shows that "Agile Projects are 3X more likely to succeed than waterfall projects. And waterfall projects are 2X more likely to fail."[3] In addition, their data indicates that "large agile projects succeed at twice the rate of non-agile projects and fail half as often."[4]

Clearly, the ability to react to market demands and deliver value with speed has become a key differentiator in digital products and companies. But it does not end there. This need for speed is not limited to the "digital" world. Nicola Accialini describes in his book *Agile Manufacturing: Strategies for Adaptive, Resilient and Sustainable Manufacturing* that there is a growing need for flexibility and speed on the production line and the ability to respond to changing markets and increasingly customized offerings, in ever shorter cycle times.[5] We recognize the growing imperative to respond to changing priorities as we build *cyber-physical systems*—that is, those that combine software, firmware, hardware, and manufacturing.

Applying Agile and DevOps ways of working in the development of cyber-physical systems through manufacturing presents an opportunity to reap significant rewards, as these systems are often large, complex, and worth millions to billions of dollars. Companies that embrace these practices can achieve speed to market and can experience increased adaptability, shorter delivery schedules, reduced development costs, increased quality, and higher transparency into delivery.

This need for speed in cyber-physical systems goes beyond the commercial industry. In the 2018 National Defense Strategy, US Secretary of Defense James Mattis called out the need to accelerate the delivery of weapons systems capabilities in response to ongoing threats and the need to ensure the continuation of national security.[6] Specifically, he highlighted "speed of relevance," as it was expressed that "the department will transition to a culture of performance and affordability that operates at the speed of relevance."[7]

Advancements in technologies and capabilities also continue to evolve quickly in the growing space market. According to Benchmark International, in 2021, the space market "was valued at $388.50 billion and is expected to reach $540.75 billion by 2026."[8] Reduced costs of developing capabilities and services with faster deployment are broadening the playing field and increasing competition. Those who don't keep up will quickly be left behind.

Left Behind

The benefits of adopting Agile ways of working are already being realized, and some organizations are anxious to reap these benefits. Even so, according to those responding to the *Competitive Advantage through DevOps* survey conducted by *Harvard Business Review*, only 10% said their organization is very successful at rapid development and deployment of software.[9] This means 90% of the market is at risk of losing out to their competition if they do not evolve and improve more quickly.

The need to change is paramount, as explained by the work of Carlotta Perez, who has been researching the different waves of technological revolutions throughout history. Her research explains how each technological wave has redefined the means of production fundamentally, which triggers an explosion of new businesses, followed by the mass extinction of those that thrived in the previous technology wave but failed to adapt.

Perez suggests that we are in the middle of the fifth technological revolution, known as the "Age of Information and Telecommunications" or the "digital age," characterized by a shift toward more decentralized and flexible production, a greater emphasis on knowledge-based services and the creative economy, and a growing concern for sustainability and social inclusion.[10]

Currently, digital environments and tools are making product development less expensive by enabling us to move physical components into the digital space and allowing product teams to perform multiple, iterative redesigns cheaply before we must acquire the needed physical components. This results in faster learning and savings in cost and schedule by finding defects earlier in the life cycle and reducing rework. Digital engineering, as it is collectively referred to, is providing the opportunity to shorten feedback loops in physical development to the speed of software through models, simulators, emulators, digital twins, and additive manufacturing.

While the new technological revolution brings clear advantages, Perez's work also shows the danger in not transforming. Based on her research, when companies continue to apply infrastructure concepts from the previous technological revolution to the current wave, they inevitably fall behind and fail, becoming casualties of each wave's mass-extinction event.

The government sector is experiencing these same challenges. The world's largest employer, the US Department of Defense, acquires cyber-physical systems in the form of strategic defense and weapons capabilities costing billions of dollars. According to a US Government Accountability Office (GAO) report, over 60% of large weapons systems are not meeting schedule and are over budget.[11] Google's CEO, Eric Schmidt, said the US government's dithering has left the country well behind China in the race to build out 5G technology.[12] China is also threatening US superiority in space. Officials of the Space Force, Defense Innovation Unit, and Air Force Research Laboratory state the United States must act quickly to maintain its advantage over Beijing, which includes using more commercial technology.[13] Based on these trends, it is imperative for the defense community to improve the rate at which they adopt Agile and DevOps ways of working so they can deliver faster than their competitors.

Consumer expectations also continue to grow, forcing both existing and new businesses to be relentless with innovation. According to Eric Ries, author of *The Lean Startup: How Today's Entrepreneurs Use Continuous Innovation to Create Radically Successful Businesses*, the only way to win is to learn faster than anyone else.[14] This includes understanding what the customer wants and delivering it faster than your competition, delivering at the speed of need.

Forbes stated that many companies have failed to meet customer expectations and are struggling to keep up with the pace of the digital age. Since 2000, approximately 52% of the companies on the Fortune 500 list have become obsolete.[15] In addition, Gallup, a global analytics firm, has written many articles regarding "quiet quitting," a trend of increasingly disengaged employees leaving the workforce at alarming rates.[16]

This creates a complex challenge: the need for innovation, the need to be able to respond and keep up with the pace of change, and the need for knowledge workers are all growing. The increasing challenge to stay relevant and meet the growing needs of the knowledge-based workforce requires companies to make culture center stage to attract and retain talent. Knowledge workers are problem solvers who have built their craft through education and hands-on experience. They seek a high level of autonomy and creativity to stay engaged, which requires a change in not only how we manage and build cyber-physical systems but how we inspire the talent that is building those systems. Moving toward Agile/DevOps ways of working helps provide an ideal working environment for knowledge workers.

Given the economic and national imperatives we face, both industry and government need to reimagine how to build, manufacture, and sustain cyber-physical systems. For one, system complexity has grown, and uncertainty is the new normal in product development. Second, the digital revolution has dramatically raised customer expectations. And finally, the workforce to power the digital economy is different from the previous one and requires new ways of working to be successful.

As organizations face these challenges, they must inspect and adapt to the world around them and continuously improve their operating model.

The software industry demonstrated its ability to respond to changing market demands and meet business outcomes using Agile/DevOps. Now it is time for the world's largest enterprises to consider how they will scale these practices across large, complex systems. There is an urgent need to iterate and deploy faster, adapt to changing needs, reduce cycle time for delivery, increase value for money, improve transparency, and leverage innovations.

According to a summary provided by Anthony Mersino of Vitality Chicago, the 2021 Standish Group *CHAOS Report* states that data from the industry continues to demonstrate that Agile projects are three times as likely to succeed as projects run with traditional project management. And waterfall projects are twice as likely to fail.[17]

What is even more interesting is when you correlate Agile adoption with the size of the project. The results of the report conclude that "large agile projects succeed at twice the rate of non-agile projects and fail half as often. 'Medium-agile' projects do not fare that much better (31% versus only 19% for non-agile projects). Only in the small category does non-agile come close to agile."[18]

PROJECT SUCCESS RATES

METHOD	SUCCESSFUL	CHALLENGED	FAILED
Agile	42%	47%	11%
Waterfall	13%	59%	28%

Figure 0.1: Agile vs. Waterfall
Source: Anthony Mersino, "Why Agile Is Better than Waterfall (Based on Standish Group CHAOS Report 2020)."

With cyber-physical systems, manufacturing brings along considerations such as designing for manufacturability, parts management, and supply-chain needs and constraints that were less pervasive in software-only systems. Leveraging tools such as digital twins supports our ability to use telemetry from the factory to optimize product development and the overall flow of value to stakeholders. Similar to product development, the practices and emergence of digital capabilities in manufacturing are also evolving rapidly. Today, as the landscape continues to shift, factories are discovering the need for adaptability *in addition to* improving flow, resulting in increased factory output, better workforce utilization, operational flexibility, reduced production costs, and increased customer satisfaction.

Bridging the Gap

Given the clear advantages of adopting Agile and DevOps ways of working in cyber-physical systems, what is preventing organizations from adopting these ways of working? After all, several companies have already shown what can be achieved with these practices. How can we bridge the gap to show that what works for software and for manufacturing can work for systems that combine the two?

The automotive industry, for instance, has been forging ahead with Lean, Agile, DevOps, and digital capabilities for years, and their experiences can be a great example of what is possible in cyber-physical systems. Many Agile principles, like having a startup mentality and the release of a minimum viable product (MVP) for rapid feedback, have been critical to the success of Porsche since the 1930s. According to Porsche AG, "In order to meet the modern requirements of volatility, uncertainty, complexity, and ambiguity, companies must become more agile (i.e., more flexible, dynamic, and interconnected)."[19] In 2021, Porsche shared how they are scaling Agile across both digital and hardware entities: "Bringing together the two worlds of physical and digital is one of our must-win battles: hardware AND software or digital products and services. Moreover, we shape the transformation with the goal of forming an attractive organization with shared values."[20]

Of course, Porsche is not alone. BMW also has demonstrated success with Agile. In a 2018 article, they highlighted specific improvements they have experienced along their journey to improve their software practices and improve delivery times.[21] The approach at BMW addressed mindset and culture, tools, and practices. As a result, their adoption:

> . . . led to improvements in the way BMW handles the development of products and has helped significantly reduce the time to market— perhaps best demonstrated in the company's new approach to interior design. Historically, Waltram's small in-house R&D team would work using physical mockups of car interiors only twice a year. Under the new approach, his team is now using software to visualize mockups in a 3D game engine, allowing for new integrations every week.[22]

In *Project to Product: How to Survive and Thrive in the Age of Digital Disruption with the Flow Framework*, author Mik Kersten highlights the BMW Group Leipzig and the actions they took to improve flow across their value stream to meet business outcomes. Not only did they address improvements from a software perspective, but they also looked more broadly at how to improve flexibility and adaptability on the production line. Their improvements created the ability to respond to increased demands more easily in the production line by adding more automation or parallelization to the given line to increase the volume of output.[23]

No discussion on Agile and DevOps in the automotive industry can happen without including Tesla. Tesla has been taking Agile adoption into the hardware space using practices often referred to as eXtreme manufacturing.[24]

For example, Tesla used these foundational principles (in particular, optimizing for change) in the construction of their factory in Shanghai in 2019, when they met their aggressive schedule and surpassed industry expectations.[25] Tesla Shanghai used digital technology to improve operational efficiencies, such as the Giga Press, a large-scale casting machine, which can create the "entire front and rear segments of the car."[26] This technological advancement "saves 300 robots, a thousand parts and 30% of the factory space."[27]

But Tesla is not done improving. There are efforts to continue optimizing the layout and operations of the factory. With ongoing investment in new technological capabilities, there is the potential to increase the speed by at least ten times and up to one hundred times. When this is achieved, it will "enable radical factory and production speed improvements."[28] These are just a few of the examples from industry on how Lean, Agile, DevOps, and digital capabilities yield positive results.

Toward a Practice of Industrial DevOps

Speed, flexibility, and adaptability are an imperative across the value stream. When we couple the results of Agile and DevOps implementation in software development with Lean and Agile in manufacturing, we have the foundational success patterns for the development, manufacturing, and deployment of cyber-physical systems in the modern age. The benefits that have been obtained across industries *can* be transferred to the cyber-physical domain, and they have the potential to provide an even greater impact on the delivery of products. This can be achieved through the application of what we have defined as *Industrial DevOps*: a set of proven principles and success patterns for building better systems faster to achieve business outcomes.

In fact, over the years, we have been exploring how to adjust Agile, Lean, and DevOps ways of working to specifically address the concerns and unique challenges of cyber-physical systems. With our over fifty years of combined experience in software and systems engineering of large, complex systems, from satellites, submarines, and spacecraft to large communications and data systems, we have witnessed the evolution of engineering and management practices. We have seen what works and what does not work. Based on these experiences and our collaboration with multiple industry working groups, we have further defined this successful application of Agile, Lean, and DevOps principles in the cyber-physical realm as *Industrial DevOps*.

Industrial DevOps is "the application of continuous delivery and DevOps principles to the development, manufacturing, deployment, and serviceability of significant cyber-physical systems to enable these programs to be more responsive to changing needs while reducing lead times."[29] The nine principles of Industrial DevOps encompass a holistic perspective, applied across the organization, and include the culture and mindsets of an organization leading to the delivery of value to the end customer.

Industrial DevOps is the advancement of Agile and DevOps practices specifically in complex, regulated cyber-physical environments. These environments contain significant challenges, considerations, and success patterns. We have learned from our decades of experience in this industry that the words of W. Edwards Deming still hold true:

> A common disease that afflicts management and government administration the world over is the impression that, 'Our problems are different.' They are different, to be sure, but the principles that will help to improve quality of product and of service are universal in nature.[30]

Industrial DevOps is the embodiment of this belief.

How to Read This Book

Throughout this book, we will show how to successfully adopt Agile/DevOps ways of working at industrial scale through the adoption of Industrial DevOps. Through nine key principles, organizations will learn how to bridge the gap between software engineering and systems engineering, how to build cyber-physical systems with speed and quality, and ultimately, how to stay ahead of their competition to deliver value to their customers.

We have organized the book into several parts. To start, Part I expands on the reasons we must change how we work, explains what Industrial DevOps is, details the nine key principles, and expands on the benefits of adopting Industrial DevOps in your organization.

In Part II, we dive deeper into each of the nine principles, where we explore the underlying concepts and how they are applied to different domains.

In Part III, we bring it all together and help show you how to start adopting these ways of working in your company or on your team. We also dig into the many barriers that currently block adopting Industrial DevOps and show you how to overcome them.

Throughout the book, you'll find case studies of real-life companies adopting these ways of working, coaching tips, questions to ask, and key takeaways.

We routinely use the example of a CubeSat (a cube-shaped miniature satellite) throughout the book to illustrate the concepts being discussed. This running example helps us build a common mental model as we discuss and explore the principles of Industrial DevOps. We chose a CubeSat specifically because it is a cyber-physical system that is relatively simple to understand but can scale in size and number to demonstrate increasing levels of complexity. If you are unfamiliar and need a crash course in CubeSats, head over to Appendix A.

Appendix B outlines the bodies of knowledge that have been combined and used to inform and design the nine key principles of Industrial DevOps. If you are unfamiliar with any of these bodies of knowledge, be sure to check out the materials in Appendix B. Often, it seems like we are all speaking different languages, but to be successful in the next phase of the industry, we must learn to speak the same language. This section will help get us there.

Appendix C summarizes the variety of tools and techniques mentioned through this book. It can serve as a reference for you and support your Industrial DevOps implementation.

Who Should Read This Book

- You're a program manager in the organization who is looking to improve delivery lead time for cyber-physical systems.

- You're a systems architect who wants to understand how Agile and DevOps can be applied to cyber-physical systems.
- You're at the executive level, sponsoring an Agile, DevOps, Lean, or digital transformation initiative across the enterprise. Our research and stories will help you understand the broader application of these principles beyond software and how it impacts the entire value chain to include all functions of the organization.
- You're sitting at the portfolio level, or you're a program manager seeking ideas to evolve and scale your existing Agile software development approach to be more inclusive of hardware teams and Lean and Agile manufacturing teams. You are seeking to improve operational efficiencies, reduce risks, and improve schedule delivery.
- You're a technical manager or leader or a plant or site manufacturing manager focused on operations, and you have noticed a gap between Agile software development teams and the other teams who also have responsibility for the design, build, and delivery of the product. You are interested in exploring new ways of working, from the ideation stage through manufacturing and delivery of products.
- You're a DevOps or Agile coach and a primary change agent for the customer or organization you serve.
- You're curious and looking for principles and ideas to apply in the development of cyber-physical systems.

This book focuses on large-scale, complex environments. However, the principles can be applied in a variety of setting and domains. We are excited to support your learning journey. Our goal is to educate and inform while providing concrete actions that can be taken to lay out a strategy and road map for improvement. This is an opportunity to learn and improve your current ways of working in the cyber-physical realm to improve business outcomes and build happier, more engaged, and empowered teams.

PART I
APPLYING THE SUCCESS OF AGILE/ DEVOPS TO CYBER- PHYSICAL SYSTEMS

CHAPTER 1

TOWARD A PRACTICE OF INDUSTRIAL DEVOPS

> Change is the law of life. And those who look only to the past or present are certain to miss the future.
>
> —John F. Kennedy

The notion of iterating and accepting "vague" requirements for large-scale, cyber-physical, and often safety-critical systems may seem unrealistic. For many cyber-physical organizations, this has been a stopping point to adopting these new ways of working. However, this is exactly how some of our greatest accomplishments have occurred. Consider the Apollo 11 mission that achieved the first human landing on the moon on July 20, 1969. The level of uncertainty throughout that mission was high, and the need to learn fast was critical. But they iterated, learned, and eventually landed on the moon.

In today's fast-paced business environment, organizations must adapt to changing market conditions, emerging technologies, and evolving customer needs to remain relevant. Agile and DevOps practices give us the means to do so. Software may have been the point of entry, but it is not the end. Agile/DevOps practices can be adjusted to fit the unique challenges and needs of building cyber-physical systems by shifting our mindsets and changing the way we work. Industrial DevOps shows us the way.

Bringing Agile/DevOps to Cyber-Physical

Since we first landed a man on the moon in 1969, digital engineering technologies have continued to mature with leaps and bounds. The cyber-physical world *can* take advantage of these digital capabilities to gain flexibility and adaptability even in the physical space, giving companies that create cyber-physical systems the ability to disrupt the market and improve delivery times.

In fact, digital capabilities create the opportunity to shift physical system development into a digital realm using tools such as emulators, simulators, digital modeling, and digital twins. Today, we also have the advantage of computer-integrated manufacturing, 3D printing, and additive materials that reduce costs over traditional methods. These abilities give cyber-physical teams greater flexibility in the design of physical products and the ability to test more frequently.

And let's not lose sight of changes happening in manufacturing. With the emergence of Industry 5.0[*] and the smart factory, how work is performed on the factory floor is also changing. The factory itself is now a cyber-physical system used to build cyber-physical systems. This is a result of the adoption of the "Internet of Things (IoT), cloud and edge computing, Artificial Intelligence (AI), big data and analytics, blockchain, robotics, drones, 3D printing, Augmented Reality (AR), and Virtual Reality (VR), Robotic Process Automation (RPA) and mobile technologies."[1] These digital capabilities help reduce the cost of traditional manufacturing methods, enabling development to frequently validate and test multiple design options and create iterative capabilities.

Additional practices such as modular hardware designs and production flexibility, robotics, automation on the factory floor, and other enabling digital capabilities are improving the speed of delivery from the ideation phase through development, production, and operations.

So, with these new technological advances, what's holding back companies from adopting new ways of working? The challenge may lie in outdated mindsets. The previous constraints of physicality have been reduced, paving the way for a new way of building.

The traditional development of cyber-physical solutions has been conducted through a serial life cycle flow of design, development, and testing (known as waterfall). The stage-gate milestones of this waterfall life cycle focus on completed documentation (and lots of it) versus validated capabilities. This results in creating a lot of documentation but a nonfunctioning system. The net result is slow time to market, lower quality, cost overruns, and solutions that do not fit their intended purpose. Industrial DevOps does not advocate for zero documentation; instead, it advocates for right-sized documentation in concert with ongoing, iterative development and validated capabilities.

---

[*] Industry 5.0 is a new and emerging phase of industrialization that describes humans working in concert with advanced technology and AI-powered robotics.

These waterfall practices have been used for decades. In the past, the cost of change was much higher, so organizations focused on controlling change to keep costs low. Today, the cost of change has diminished thanks to technological advancements. Today the risk of *not* changing is the bigger monster. Jeanne W. Ross of the MIT Sloan Center for Information Systems Research stated, "Clearly, the thing that's transforming is not the technology—it's the technology transforming you."[2]

Digital transformation has also increased customer access to information, which has skyrocketed customer expectations in everything from innovation to the speed of delivery through operations. These expectations are prevalent for digital and cyber-physical products, such as smartphones, wearables, smart appliances, medical devices, vehicles, and even weapons systems.

According to McKinsey, the COVID-19 pandemic further accelerated the digitization of companies and their supply chains by three to four years, with some digitally enabled products accelerated by seven years.[3] It has never been clearer that now is the time for companies that build cyber-physical systems to adopt new ways of working.

Applying the theory, practice, and learnings from Agile and DevOps has the potential to dramatically improve the development and delivery of cyber-physical systems. Companies that solve this problem will increase transparency, reduce cycle time, increase value for money, and innovate faster. Simply, they will build better systems faster, and they will become the ultimate economic and value delivery winners in the marketplace. These practices are especially useful as the systems become increasingly complex with growth in unintended emergent behaviors.

Taking proven principles and practices from Lean, Agile, and DevOps and implementing them at the system level with a common language and mental model is actually a simple idea. That simple idea executed well can cause world-changing ripples in product development.

In 2012, I (Robin) had the opportunity to support fighter jet teams to deliver updates to a legacy cyber-physical system with multiple safety requirements. We had an aggressive but necessary schedule of thirteen months to deliver the updates. Given the schedule constraints, leadership was willing to take a risk on a new approach to development.

I coached an initially reluctant set of teams through an Agile transformation. The teams leveraged a tiered planning approach, decomposing their work by product, working in pairs, developing in timeboxes, holding daily stand-ups, and performing demonstrations of their work. At the close of each time box (sprint), the teams held a retrospective and identified one change they could apply to the next sprint.

The impact the Agile approach had on the system was immense. The system was completed in seven months as opposed to thirteen, with a record-low number of defects in hardware integration. The impact that the Agile approach had on the team and their morale was beyond amazing. Aerospace engineers, who had been building and maintaining aircraft for over thirty years, claimed they had never had more fun during the development process or a greater impact on the system.

As organizations realize the benefits of Industrial DevOps, the opportunity to disrupt the status quo of the entire product life cycle presents itself. For the US Department of Defense (DoD), this is imperative in ensuring the safety and freedom of the United States and its allies. For the space industry, it means getting the edge on space advancements and human discovery. For the broader community, it provides an opportunity to outpace their competition.

According to the National Science Foundation (NSF), "Advances in [cyber-physical systems] will enable capability, adaptability, scalability, resiliency, safety, security, and usability that will expand the horizons of these critical systems. [Cyber-physical system] technologies are transforming the way people interact with engineered systems, just as the Internet has transformed the way people interact with information."[4]

Despite the clear evidence that Agile and DevOps practices have played a key role in the success of software development organizations, many organizations that build cyber-physical systems have the mistaken idea that their systems are too complex to use Agile or DevOps practices. From experience, we recognize that even the simplest of ideas can be difficult to implement when working against cultural norms. What we need is to look at the problem we have been solving from a different perspective. Henry Ford is credited with saying, "If I had asked people what they wanted, they would have said faster horses." People did not consider that an entirely new form of transportation could be made available.

Current State of Cyber-Physical Systems

The term *cyber-physical system* was first used in 2006 by Helen Gill at the US National Science Foundation. According to the NSF, cyber-physical systems "integrate sensing, computation, control and networking into physical objects and infrastructure, connecting them to the Internet and to each other."[5] Building from this definition, we include systems that are part of private, secure networks and communications infrastructures. These systems, including software, hardware, and manufacturing components, are often complex and costly to build. Many cyber-physical systems have safety and security requirements, making these systems even more challenging to adapt to changing priorities and technologies: "Security threats have a high possibility of affecting [cyber-physical systems, and they] can be affected by several cyberattacks without providing any indication of failure."[6]

Cyber-physical systems are everywhere. You see and use these systems as part of your daily activities. They exist across industries in many different forms. Cyber-physical systems can be found in the automotive industry, agriculture and farming, aeronautics and space systems, undersea systems, energy systems, medical and health care, communication devices, smart factories, smart grids, wearable devices, and more.

While there are challenges to the adoption of Agile and DevOps in cyber-physical systems, it is far from impossible. Let's look at some early adopters of Industrial DevOps principles and practices.

The Early Adopters Are Reaping Success

Industrial DevOps practices provide an innovative approach to address many of the unique challenges faced by the development and manufacturing of cyber-physical systems. Where we see pockets of these principles being adopted, the companies are recognizing clear benefits. Early successes of some Industrial DevOps principles can be found in companies such as Tesla, SpaceX, Planet Labs, Bosch, and Saab Aeronautics.

Tesla

To build a competitive presence in the auto industry, Tesla applies what they call "first principles" coupled with attention to how people collaborate. This means when they want to improve a system and innovate faster, they take a systems view. Innovative thinking increases through increased knowledge sharing and strong team collaboration.

Through the efforts of Joe Justice, chair of the Agile Business Institute (ABI) and CEO of Wikispeed Inc., applying an Agile mindset and practices has been fundamental to Tesla's ways of working and their success in building a continuous improvement culture. Tesla CEO Elon Musk has stated the importance of innovation at Tesla: "Speed of innovation is what matters."[7] As reported in the ABI article "Tesla Agile Success," while innovations are ongoing at Tesla, some specific innovations include "designing the battery casing as part of the car structure, the Giga Press, the octovalve and its manifold, and enhancing their car seats from being the worst in the early Model S to being the best on the market (according to Sandy Munro)."[8]

One well-known investment is the Giga Press. The use of the Giga Press to manufacture large parts of vehicles has resulted in increased speed of production and improved operational efficiencies. By using the Giga Press to manufacture parts for the Model Y, the Fremont factory was able to reduce the body shop by 30%, or about three hundred fewer robots, as compared with the manufacturing of the Model 3.[9] Taking a systems view and building teams for improved collaboration helps Tesla build better systems faster. Today, they have a larger market cap than the next six auto companies combined.[10]

SpaceX

Anyone who is familiar with Agile principles and is watching SpaceX recognizes the importance these principles play in the innovative learning cycles of their space systems. As reported in the article "SpaceX's Use of Agile Methods," SpaceX makes vast use of end-to-end systems modeling, and as Elon Musk says, they can "take the concept from your mind, translate that into a 3D object, really intuitively . . . and be able to make it real just by printing it."[11]

The Industrial DevOps principles have been demonstrated by SpaceX through their ability to test often and early and get rapid feedback on their testing. Using these principles, in December 2022, "SpaceX has now edged out Lockheed, which has a valuation of $137 billion, making it the third-most-valuable aerospace and defense franchise in the western world behind only Raytheon Technologies (RTX), which holds the top spot, and Boeing (BA)."[12]

Planet Labs

Planet Labs is an Earth-imaging company that operates a fleet of over two hundred satellites, providing high-resolution imaging of the Earth's surface for a variety of stakeholders, from agriculture to government to energy. Planet Labs refers to their approach as "Agile aerospace," where, over the last decade, they have completed fourteen major iterations of the Dove spacecraft design. Planet Labs explains that getting their first satellites into space quickly enabled them to learn many lessons about constellation management and optical systems before the cost of change was too expensive. These days, Planet Labs launches satellites into space every three to four months. They were able to parlay early learning into a company that is currently estimated to be valued at over $2.8 billion.[13]

Bosch

Bosch is a German engineering and technology company whose products are seen in our everyday life. For several years, the organization has been leveraging Agile and DevOps principles across several areas, such as their autonomous systems and control units, sensors, smart heating controls, and power tool division. Soraia Ferreira, a software engineer supporting Bosch in the area of smart heating and heating systems, stated that by applying Agile practices, "we created a product that has been a smash hit in the market," and "without innovative physical heating technology, we wouldn't make any progress with IoT software."[14]

Saab Aeronautics

The Gripen fighter jet, owned by Saab Aeronautics, has been one of the earliest adopters of Agile, Lean, and Scrum for a large cyber-physical system. For the JAS 39E Saab Gripen, there were two thousand to four thousand individuals working across every level and across software, hardware, and the fuselage. Based on the article "Owning the Sky with Agile: Building a Jet Fighter Faster, Cheaper, Better with Scrum," Agile practices have enabled Saab "to manage variability and drive performance with clarity and commitment. The result is an aircraft delivered for lower cost, with higher speed, and greater quality."[15]

NASA

NASA has some of the world's largest cyber-physical systems in the space domain. They, too, have embraced Agile, Lean, and DevOps principles across some parts of the agency. The Federal News Network captures how their Agile implementation has helped align strategy to execution, with the ability to deliver value incrementally with fast feedback from stakeholders through regular system-level demonstrations. As quoted in the article "Securing Containerized Applications," Shenandoah Speers, NASA's director of application and platform services in the office of the CIO, stated that his team has "created a DevSecOps pipeline platform that allows them to do on-demand continuous integration (CI) and continuous deployment (CD) utilizing containerization to automate the build security, scanning and deployment process."[16]

———

Wider adoption of Agile and DevOps principles continues to expand into other functions, as demonstrated through industry professional organizations such as the Project Management Institute, International Council on Systems Engineering (INCOSE), National Defense Industrial Association, and Software Engineering Institute. For example, in June 2022, at the INCOSE 32nd Annual International Symposium, a working session, "SE Modernization Strategy Session Follow-Up," was held to discuss systems engineering modernization to include Agile practices, and the organization continues to promote the practice Agile Systems Engineering. These organizations are taking positive steps to include Agile and DevOps into areas beyond software development.

A 2022 report from the US Government Accountability Office (GAO) highlights the DoD's emphasis on iterative development and continued modernization of its software development efforts. While iterative development and software modernization are on the rise, fast feedback and the frequent release of capabilities for these systems is still suboptimal. The GAO has also reported on the significance of cybersecurity for weapons systems. These capabilities are critical for ensuring the safety and security of freedom.[17]

These examples are not meant to imply that the efforts of these organizations are perfect; however, it is clear that these organizations are experiencing positive results through a set of Lean and Agile principles for the development and production of cyber-physical systems. Companies that solve this problem first will increase transparency, reduce cycle time, increase value for money, and innovate faster. **They will build better systems faster and become the ultimate economic and value delivery winners in the marketplace.** It is evident that we can leverage Agile and DevOps practices to address the unique challenges of cyber-physical systems and obtain similar outcomes.

What Is Industrial DevOps?

Agile and DevOps have been implemented in software for over a decade. This is not a new phenomenon, but adoption in different environments beyond software is still evolving, as we have explored here and in the Introduction. The solution for wider adoption of Agile/DevOps in cyber-physical systems requires a unique set of principles, which we have coined *Industrial DevOps*.

Industrial DevOps is the application of Lean, Agile, and DevOps principles to the planning, development, manufacturing, deployment, and serviceability of significant cyber-physical systems. The practice of Industrial DevOps pulls from multiple bodies of knowledge, including Agile, Lean, DevOps, and systems thinking, as well as from our own personal experience delivering cyber-physical systems in the new technological revolution.

Industrial DevOps bridges the principles and practices of Agile, Lean, and DevOps with the unique needs and challenges of cyber-physical systems through nine principles.* By doing so, Industrial DevOps enables organizations building cyber-physical systems to be more responsive to changing priorities and market needs while also reducing lead times and costs. Through research and experience, we have found that the combined use of these nine principles is effective in successfully delivering cyber-physical systems across industries.

1. **Organize for the Flow of Value:** Organizing for flow provides guidance on how to align your multiple product teams for regular demonstration and delivery of value.
2. **Apply Multiple Horizons of Planning:** Apply multiple horizons of planning to address scaling and complexity while leveraging ongoing experimentation and learning.
3. **Implement Data-Driven Decisions:** Data-driven decisions use current observations and metrics to determine the state, manage the flow of work across systems of systems, and continuously improve with real-time data.
4. **Architect for Change and Speed:** Architecting for change and speed provides information on multiple architecture considerations, which can reduce dependencies and improve the speed of change.
5. **Iterate, Manage Queues, Create Flow:** Iterate, manage queues, and create flow to emphasize the importance of fast feedback, experimentation, and continuous learning.

---

* These principles are a modification from the original eight defined in our earlier publication, *Industrial DevOps: Applying DevOps and Continuous Delivery to Significant Cyber-Physical Systems.* The original principles were created with input from multiple contributors from a wide variety of companies across different industries. The nine principles in this book still align with those original principles but have been refined to provide more clarity as we have continued to learn from new experiences and apply a growth mindset to all that we do, which led us to the additional ninth principle: Apply a Growth Mindset. We are committed to continuously learning and growing in our experiences and in our thinking.

6. **Establish Cadence and Synchronization for Flow:** Establishing cadence and synchronization discusses how these two concepts complement each other to reduce variability and improve predictability.

7. **Integrate Early and Often:** Integrating early and often covers different levels and types of integration points across large, complex systems.

8. **Shift Left:** Shifting left emphasizes a "test-first" mindset encompassing the multiple levels of testing across cyber-physical systems.

9. **Apply a Growth Mindset:** Applying a growth mindset expresses the need to continuously learn, innovate, and adapt to the changes around us in order to stay competitive.

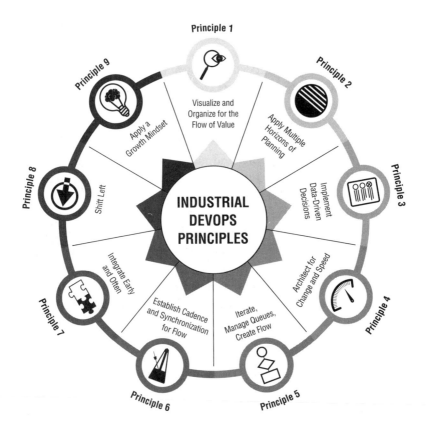

Figure 1.1: Industrial DevOps Principles

Let's take a look at Industrial DevOps in action through a brief glimpse into each of the nine principles. We will take a deeper look into each principle in Part II of this book.

1. Organize for the Flow of Value

The first step in adopting Industrial DevOps is to visualize and organize around the flow of value instead of around functions. This may sound obvious, but many companies are organized around functional activities, such as systems engineering, hardware engineering, software engineering, test engineering, etc. This type of organization creates multiple hand-offs and lots of documentation. Instead, teams should be organized around value streams, and the teams building the systems within the value stream include people with all the skills needed to improve the flow of value and shorten delivery cycle. (See Chapter 4 for more.)

2. Apply Multiple Horizons of Planning

Predictive planning with phase gates has been the most popular approach to building cyber-physical systems. The belief has been that short-term empirical planning, which allows software systems to design and iterate at speed, won't work for cyber-physical systems, with their longer lead times for hardware, dependencies across systems and systems of systems, regulatory controls, and more.

Industrial DevOps employs *multiple horizons of planning* to address this unique challenge. This helps move organizations away from long predictive planning to the short-term planning common with Agile. This approach allows teams to obtain empirical data quickly from planning horizons and apply the knowledge to the next planning horizon, always adjusting the planning based on empirical data. (See Chapter 5 for more.)

3. Implement Data-Driven Decisions

The third principle of Industrial DevOps focuses on using empirical data and leading indicators to better understand the progress and state of the product we are building. The data is then used as input into the decision-making process as the next cycle of work is planned and prioritized. Using data to drive decisions also provides the ability to measure the results of those decisions. (See Chapter 6 for more.)

4. Architect for Change and Speed

Architecture is a critical element in the ability to deliver products and services at speed. Concurrent development is much faster than synchronous development. In Principle 3, we decomposed our systems into smaller components and threads that can be validated and verified, allowing us to learn faster. In Principle 4, we apply a modular architecture to address speed and change, as it enables small teams to build and deliver faster with reduced dependencies between other teams who may be working on the same system module.

Through modularity, a team can change part of the design without impacting the other parts of the system. Architecting the system for serviceability means considering how the system components are updated or enhanced during the ongoing development of the system, whether before or after deployment. This approach makes change easier and increases the speed of delivery. (See Chapter 7 for more.)

5. Iterate, Manage Queues, Create Flow

In Principle 1, we organized our teams for flow. Now, in Principle 5, we create and maintain that flow using iterative and incremental approaches to build out complex solutions. This principle couples Agile's approach of iterative and incremental development with Lean's commitment to reducing batch size to increase flow. (See Chapter 8 for more.)

6. Establish Cadence and Synchronization for Flow

In Principle 6, cadence and synchronization come together to improve flow and establish predictability, which is critical when building cyber-physical systems. A lack of predictability is one of the leading detractors for organizations to adopt more Agile ways of working.

Large cyber-physical systems typically have multiple teams designing, implementing, and deploying multiple interconnected subsystems and components over long periods of time. All of this results in unknowns and variability.

In product development, our goal is to exploit good variability and remove bad variability. The trick is knowing which is good and which is bad. Cadence and synchronization are two tools that aid in removing bad variability while providing the opportunity to exploit good variability. (See Chapter 9 for more.)

7. Iterate, Manage Queues, Create Flow

The more frequently we integrate, the faster we learn and the better our product. However, continuous integration may not be possible for every system. As we scale beyond software, we must consider constraints associated with physical products. For cyber-physical systems, our goal is to integrate as frequently as possible to evolve the system to meet customer needs and get feedback on what is working and where improvements can be made. Some considerations that impact integration are investment of automation, lead times in hardware, expensive test equipment, regulatory compliance rules, and training of employees on the tools and processes. (See Chapter 10 for more.)

8. Shift Left

We must begin with a clear idea of how we are going to test so we can build quality into our products and services. This means thinking first about how we will test so that we have achieved the desired outcome before we start the development of products and services. The importance of early integration and iterative testing is not uncommon in the hardware domain. Models, prototypes, and simulations have been adopted for years. We now align these practices with being test driven, developing iteratively, and organizing around the flow and delivery of value. (See Chapter 11 for more.)

9. Apply a Growth Mindset

Finally, whenever we undertake dramatic change, we must ensure that we approach that change with a growth mindset. Without this final principle in place, none of the other eight matter. This is the principle that glues all the others together. We will discuss what a growth mindset is, why a growth mindset is important, and how to build a growth mindset both individually and as an organization. In addition, we describe the relationship to successfully driving toward Industrial DevOps principles by leveraging the power of a growth mindset. (See Chapter 12 for more.)

Conclusion

The principles of Industrial DevOps have been built from the integration of existing principles. Just like all the innovators before us, we have taken some good ideas and applied them in new ways that we have found to work successfully for our unique needs. Our goal is to provide you with some insights to help you jump-start your approach to delivering products and services better. In the next chapters, we look at the key benefits and misconceptions of applying Industrial DevOps to the development and production of cyber-physical systems.

KEY TAKEAWAYS

- There are companies applying Lean, Agile, and DevOps who are seeing positive results. We can leverage these successes to build more success.
- Industrial DevOps principles extend across the value stream at all levels of the organization.
- Industrial DevOps is a new way to look at building systems of systems with hardware, software, and firmware/manufacturing.
- Benefits of Industrial DevOps include shorter lead times, improved transparency and visibility of progress, increased predictability, improved quality, reduced cost, and improved morale.

QUESTIONS FOR YOUR TEAM TO ANSWER

- Why is it important to extend the Industrial DevOps principles beyond software and across the functional areas?
- Which principles are already being applied in your team environment? Are there any principles being applied at scale in your organization?
- Are there principles that are easier to apply in software than hardware? If so, why? What would your team need to drive the principles into hardware development?

COACHING TIPS

- Industrial DevOps principles are founded upon well-established, understood bodies of knowledge (see Appendix B). Your organization is already applying some of these practices. Use that as a starting point and improve from there.
- Consider which of the principles your organization is struggling with the most or which principles you are most curious about. Write them down and be prepared to reflect on this as you read and learn more in later chapters.

- When applying Industrial DevOps principles to cyber-physical systems, the goal is to define and demonstrate what we can learn instead of what functionality we can deliver. What is something you have recently done at work that you could demonstrate to someone on your team for feedback in the spirit of continuous improvement?

CHAPTER 2

BENEFITS OF INDUSTRIAL DEVOPS

Working with customers, leaders, and many teams as they explore the idea of adopting Industrial DevOps practices, we are almost always asked, "Why should we change?" and "What benefits should we expect?" These are valid questions, as change takes time, investment, and energy. It means questioning how work is currently being done, the underlying assumptions and beliefs of our current business operating system, and if there might be a better way. With the increased advancement of digital capabilities, a general shift to modern ways of working, and new technologies emerging regularly, leaders must continuously prioritize the investments they make against the anticipated benefits.

Lean has existed in manufacturing for decades. Nearly every company developing software has embraced some level of Agile and DevOps practices. Industrial DevOps brings these together into the development of cyber-physical systems. The integration of these practices brings together a set of demonstrated and desired benefits to build better systems faster.

In this chapter, we will introduce you to the benefits of Industrial DevOps. Based on your experiences, you may have already encountered some of the misconceptions, but for some of you, this may be new. Either way, you will have additional information to carry with you as you work with others along this journey.

Benefits of Industrial DevOps

There are multiple benefits that have been realized by companies that have implemented some or all of these practices. The goal of adopting Industrial DevOps principles is to achieve desired business outcomes. As you embark on your Industrial DevOps journey, start by understanding what benefits (business outcomes) you are seeking and how Industrial DevOps can help your organization achieve those benefits.

The benefits that most organizations are seeking fall into one or more of the following areas:

- Learn faster
- Time to market/speed
- Productivity
- Quality
- Employee happiness
- Customer satisfaction

Learn Faster

Learning faster allows us to remain competitive. As discussed in Perez's work on historical patterns of innovation and their impact on economic growth, we are experiencing the fifth technical revolution, which is referred to as the age of information and technology that began in the late 1970s. According to Innosight in 2016, 50% of the companies listed in the S&P 500 index would be replaced over the next ten years.[1] The companies that learn the fastest are likely to remain relevant. We cannot predict what products customers are going to need in the next ten years, but leveraging the short learning cycles that Agile and DevOps provide can keep us on point.

Time to Market/Speed

Time to market means getting your product out to your users/customers in the time they need it. You may be working to release just before a major holiday in the hope of increased online shopping sales, or maybe you are working to outpace your nearest competitor and get a new product released ahead of them. If you are in defense, you may be responding to increasing threats or cybersecurity needs and need to respond quickly. Regardless of why speed is important to your organization, Industrial DevOps principles can help you.

According to the *Harvard Business Review* research report *Competitive Advantage through DevOps*, in 2019, 86% of those surveyed responded that releasing new software quickly is important to outpace their competition. And 70% responded that DevOps is contributing to increased speed to market.[2] Within the automotive industry, there have been similar improvements in faster delivery, with reports of Tesla being able to deliver one-hundred-times-faster factories or production by applying Agile methodologies to hardware manufacturing.[3]

The first scenario could mean faster delivery time of new software functionality to an existing product in the field. This could mean pushing out new software features to cars that are already purchased or updates to satellite systems. Another scenario is when the software and hardware for a new cyber-physical system are still being developed. Faster time to market in this case could mean improving the overall lead time for the development of a new satellite, rocket booster, medical device, or robot for the manufacturing plant in which there is development for both hardware and software, resulting in a much longer lead time than in the first scenario.

Regardless of the scenario, the goal is to achieve faster time to market to meet specific business outcomes. Faster time to market helps us gain a competitive advantage, see the return on investments, and shorten the lead time to deliver value to the customer.

Productivity

Continuous improvement in practices and digital capabilities improves productivity, which is the ability to produce more by reducing waste and streamlining wait times. With the application of Lean, Agile, and DevOps, there has been evidence of improved productivity and operational efficiencies, which is clearly a good thing.

Based on work by Gary Gruver in 2015 while he was working at Hewlett Packard during the early adoption of the Agile transformation, they experienced significant improvements. The organization recognized a reduction in development costs ($100 million to $155 million) with a 140% increase in the number of products supported and increased capacity for innovation from 5% to 40%. What is also interesting is that the greatest improvements witnessed were not at the team level but with teams of teams coming together through regular integration in production-like environments. This resulted in greater impacts through overall improved productivity of the organization to meet business outcomes.[4]

While the benefits of Agile are often considered to be team level, research in 2021 by Paula de Oliveria Santos and colleagues uncovered the benefits of large-scale Agile. Their research analyzed seventy-six articles that highlighted Agile's recognized benefits. From that research, they discovered over thirty benefits of large-scale Agile, with one of those benefits being improved productivity. This finding was based on the implementation of multiple Agile frameworks. Their research identified that of the benefits realized, productivity was often identified as a benefit, and "adopters of Scaled Agile Framework have reported significant improvement in terms of productivity and quality."[5]

Tesla also has their own results to tout. According to an article published by Nikkei Asia, by the end of 2022, Tesla's methods replaced Toyota's Lean and produced a more profitable car. Tesla achieved $9,570 net profit per car, which is approximately eight times the profit of Toyota.[6] We recognize there are multiple factors impacting profitability and not process alone.

Improved efficiencies in one area are not the whole story. Without understanding where the bottlenecks are in the whole value stream, improving efficiencies and producing more product parts or features before a bottleneck can lead to too much inventory or wait time until the bottleneck is removed. This is exemplified in *The Goal* by Eliyahu Goldratt. Improve productivity with the entire value stream in mind and focus on removing bottlenecks so value can be delivered.

In *Sooner, Safer, Happier: Antipatterns and Patterns for Business Agility*, Jonathan Smart recognizes the shift in mindset from a focus on localized improvements in productivity to the end-to-end flow or lead time. While improvement in processes can lead to improved productivity, caution needs to be exercised. With an increase in productivity, you may be improving time to market and delivery of product features; however, it is important to ensure what you are delivering is providing *value* to the customer. He refers to this concept as "valuetivity"—that is, "the soonest realization of the most value with the least output."[7]

Improving productivity and operational efficiencies is important, but you must also look across the entire value stream to ensure improvements are made at the right place in the value stream. Improve productivity. Improve time to market. Ensure improvement in "valuetivity."

Quality

A quality product is about both ensuring we build the right product and we build the product right. Building the right product means that it provides "valuetivity." The product is working as intended, and the customer is receiving the intended value. The other perspective is that the product is built right. This means the product or its subcomponents are working and have been tested. These two concepts demonstrate the difference between system validation and verification.

As we improve the quality and value of the product, we are continuously improving the process. The improved quality of the process ties back to the benefit of improved productivity. For decades, the Lean community has touted the need for getting it right the first time. This concept means building quality in along the way versus bolting it on at the end. Teams continuously improve the product by improving their technical capabilities and receiving regular feedback from the customer.

This is also true for the cyber-physical system community. Due to the safety and security aspect of cyber-physical systems, along with the often-enormous costs of the systems, getting it right the first time (at launch) is often a requirement. However, we can do that only by building in quality with each iteration of work. Based on reports published by *Harvard Business Review*, quality continues to be a driver for the adoption of DevOps. According to their market research, 64% of the respondents claimed service/product quality benefits.[8]

The DevOps revolution has had a major emphasis on improved quality. Through improved product quality, systems are more sustainable. Defects are discovered earlier in the product life cycle, with the implementation of short iterations creating savings in cost and schedule.

According to Jeff Sutherland in *Scrum: The Art of Doing Twice the Work in Half the Time*, if a developer entered a bug into the system and fixed it right away, it might take only an hour to fix. However, if the defect was discovered three weeks later, it took at least twenty-four times longer to fix.[9] That was a software team who specialized in the development of PDAs. Imagine scaling that scenario into a cyber-physical system that has hundreds, if not thousands, of developers. The savings in time, cost, and frustration has the potential to be significant as we scale these practices.

Employee Happiness

One of the core Lean principles focuses on respect for people. It is the people who do all the work, and through their daily efforts and dedication, they are delivering results. Improving business performance and delivering value to the customer in the shortest lead time hinges on the engagement and collaboration of the workforce to deliver results.

Workplaces need to harness the involvement and enthusiasm of their employees to build better systems faster with more innovative solutions for their customers. Environments that are psychologically safe and where people are given the tools to do their job lead to increased performance. A place where people are able to experiment and innovate, where their products are used, leads to improved "employee satisfaction."[10] This creates a better work environment where collaboration and morale are high.

Lean and Agile teams are actively engaged in planning and demonstrating their team's products. Through planning across multiple horizons, they can see how their work ties back to the strategy. This creates a sense of alignment with the organization.

In 2016, *Harvard Business Review* published the article "Embracing Agile." The authors emphasize that Agile has not only revolutionized the world of software development but is also impacting how organizations lead people: "Compared with traditional management approaches, agile offers a number of major benefits, all of which have been studied and documented. It increases team productivity and employee satisfaction."[11]

Also supporting this claim is the Business Agility Institute. In 2019, they conducted research investigating employee engagement and found a positive and higher correlation between employee engagement and mature Agile organizations.[12]

Agile is transforming the world of work. It is transforming how we build systems, how we lead people, and how we work together. This shift is leading to higher levels of employee engagement, which leads to happier employees and improved business outcomes.

In 2021, *Forbes* published an article, based on data from Gallup's *State of the American Workforce* report, that says"

> Companies with an engaged workforce are 21% more profitable. Companies that lead in customer experience have 60% more engaged employees. *And study after study has shown that investing in employee experience impacts the customer experience and can generate a high ROI for the company.* And here's our favorite stat of all, proving once and for all that if you're not taking your employees' experience into consideration, your customers will go elsewhere: Companies with highly engaged employees outperform their competitors by 147%.[13]

Based on the data from the collective research, it is clear that Agile ways of working lead to increased employee engagement, and building this positive experience for your employees produces even better results for the company.

Customer Satisfaction

Our goal is to deliver value in the shortest lead time through an empowered and engaged workforce. Delivering value means demonstrating value to the customer. Industrial DevOps principles enable large development teams to continuously improve and adapt to customers' changing needs and priorities and to allow for the opportunity for increased innovation. Short, iterative development cycles provide the opportunity for increased customer engagement and ongoing feedback to the teams as the product is developed.

A 2016 *Harvard Business Review* article confirmed that based on interviewees' experience, "Agile improves customer engagement and satisfaction, brings the most valuable products and features to market faster and more predictably, and reduces risk."[14]

Research on the benefits of Agile manufacturing for smart factories has also indicated improvements in customer satisfaction because of improved product quality and the ability to meet schedule demands.[15]

Engagement with the customer, as the product is developed, helps ensure the highest-priority items are being addressed. Customer satisfaction increases as a result of improved engagement and open collaboration.

KEY TAKEAWAYS

- Most businesses have applied elements of Agile and DevOps at the software level. Now these principles are being applied in cyber-physical systems.
- Applying Industrial DevOps principles provides multiple benefits. As an organization, know which benefits you are seeking and how you will measure improvement.

QUESTIONS FOR YOUR TEAM TO ASK

- Which benefit(s) are most important to your organization?
- What is the cost/benefit or break-even point for your organization?
- How will you measure improvements? Most of the benefits are lagging indicators and take longer to be realized. How will you measure interim progress?

COACHING TIPS

- Begin with the outcome or benefit you are seeking. Capture why this benefit is important to the organization. Gather data to help communicate why this is important.
- As you consider the benefits, capture your organization's strengths, weaknesses, opportunities, and threats.
- Perform a cost/benefit assessment.
- Capture your initial baseline data to objectively measure benefits.

CHAPTER 3

MISCONCEPTIONS ABOUT INDUSTRIAL DEVOPS

As Industrial DevOps principles continue to scale to cyber-physical systems, several misconceptions have evolved. The following misconceptions, while not all-inclusive, are ones we have encountered regularly while working in cyber-physical system environments; some are also prevalent in software-centric environments, but the misconception is magnified in cyber-physical due to the typical scale. With each misconception, we provide a response and identify which chapter you can read for additional information to help understand the principle to overcome the misconception.

Table 3.1: Misconceptions about Industrial DevOps

Misconceptions about Industrial DevOps
Agile/DevOps development efforts don't plan.
Agile/DevOps programs constantly change, and that doesn't work for hardware.
Agile/DevOps does not have systems engineering practices.
Agile/DevOps programs sacrifice quality for speed.
Agile/DevOps does not have any documentation.
Agile/DevOps is only for teams, not managers/leaders.
Agile/DevOps requires deploying operations continuously.
Agile/DevOps requires everyone to be colocated.
Agile/DevOps practices are only for software.
Agile/DevOps does not work with safety-critical systems.
Agile/DevOps requires you to complete a whole system in two weeks.

MISCONCEPTION: Agile/DevOps Development Efforts Don't Plan

Planning is necessary for any product development effort. When Agile first started, it was implemented on smaller efforts and gave the perception that longer-term planning was not necessary. This couldn't be further from the truth. A general lack of experience and training (skills and knowledge) regarding Agile often leads to poor implementations, where teams mistakenly believe Agile means no planning. In some instances, teams have looked at frameworks like Scrum and thought that they needed to plan only "sprint by sprint," with no longer-term vision.

This lack of understanding has added to the larger misconception that Agile teams don't plan and has hampered the potential that Agile and DevOps offer for large efforts, specifically those with hardware, because of the long lead times. Interestingly, when using an Agile approach, teams often feel like they do more planning than they did with a traditional product development approach. In some cases, this is likely true. Agile teams plan more. With traditional approaches, the manager did most of the planning and gave it to the workers to implement. With Agile working methods, the manager provides a vision and road map, but the teams do the detailed planning for the iterations and daily work. They focus on how it gets done.

For cyber-physical systems, looking at multiple horizons of planning (Industrial DevOps Principle 2) is especially important for the alignment and integration of work across teams and suppliers. The difference is that longer horizons are kept at a higher level of granularity and are further refined as the teams get closer to implementation. The longer horizons provide a vision and forecast of what is coming in terms of features. Whereas the daily planning horizon provides the team the opportunity to shift plans as often as necessary. For example, at Tesla, the planning never stops. Teams have the autonomy for real-time changes in their plans with budget reallocation in real time every day with AI guidance, called Digital Self-Management.

To read more about planning and dispelling this misconception, read Chapter 5: Apply Multiple Horizons of Planning.

MISCONCEPTION: Agile/DevOps Programs Constantly Change, and That Doesn't Work for Hardware

Another misconception is that Agile/DevOps efforts are always changing the requirements, and that won't work for cyber-physical systems because of the hardware components. Principle 2 of the Agile Manifesto does state that we "welcome changing requirements, even late in development"; however, there are underlying premises to consider.

First, there are multiple levels of requirements and functionality that are decomposed to create a backlog of what needs to be done to build the system. It is true that this backlog of work is constantly refined and can be reprioritized. At the lowest levels of planning, the backlog will change more frequently as teams learn more about the complex system they are building. The higher-level requirements are inclined to change less frequently, especially requirements related to size, weight, power, and cost. Innovations will emerge and impact requirement definition and prioritization.

In the paper "Overcoming Barriers to Industrial DevOps: Working with the Hardware-Engineering Community," we highlighted the fact that there are physical constraints when developing cyber-physical systems. One approach to address these constraints is to "look beyond single, specialized models and use systems thinking to build an integrated set of models that allow change of parameters in size, weight, power, and cost as we test solutions and learn by applying short development iterations."[1] The adoption of digital capabilities, tools, and modular architecture makes change less costly and makes it easier to respond to change.

To learn more about planning and refinement, read Chapter 5: Apply Multiple Horizons of Planning and Chapter 7: Architect for Change and Speed.

MISCONCEPTION: Agile/DevOps Does Not Have Systems Engineering Practices

Yes, we need systems engineering practices; however, what that looks like and how it is implemented have evolved over the past couple of decades. In small development efforts, systems engineering work was most likely consumed by the development team, making that work less visible. As you scale and get into developing larger systems and systems of systems, there is a greater need for systems engineering practices. It is recognized that the practices found within the traditional Vee model are not hand-offs from one functional area to another. They are much more integrated, and as a result, these practices have been interwoven into all of the Industrial DevOps principles.

Read more about systems engineering foundations in Appendix B: Industrial DevOps Bodies of Knowledge.

MISCONCEPTION: Agile/DevOps Programs Sacrifice Quality for Speed

Implementing Agile and DevOps practices improves quality. If you're not improving quality, you are doing it wrong. Based on Principle 9 of the Agile Manifesto, "Continuous attention to technical excellence and good design enhances agility."[2] Technical excellence means always looking at the quality of your product from code complexity, automated testing, architecture, improving flow, and regular user feedback. Quality is built in. And speed and time to market require quality. We learned from the Lean community that small batch sizes can help improve quality and reduce overall costs, where defects can be found earlier as small batches flow through the system faster and users have more opportunity to engage.

To learn more about this, read Chapter 8: Iterate, Manage Queues, Create Flow.

MISCONCEPTION: Agile/DevOps Does Not Have Any Documentation

This is a frequent misconception that has existed for many years, regardless of the product or size of the effort. Part of this misconception stems from not fully reading the Agile Manifesto. It may also stem from frustrations when, early on, documentation was the primary product being delivered via traditional waterfall ways of working. The Agile Manifesto states, "We value working software over comprehensive documentation." Yes! That is so true. However, if you read on, it says, "While there is value in the items on the right [documentation], we value the items on the left more [working software]."[3]

Some level of documentation is necessary. How much documentation and the form of documentation can vary based on the system you are building. If you were to look at the documentation for a phone app versus the documentation for a missile or for the next spacecraft, the amount and form of documentation would look different. The important thing is that we don't measure progress on how much documentation has been written, but by what we can demonstrate in terms of product so we can get feedback and learn from it. Understand what level of documentation is needed for your system and how to use digital tools to ensure maintainability as the system evolves.

More on this concept is discussed in Chapter 6: Implement Data-Driven Decisions.

MISCONCEPTION: Agile/DevOps Is Only for Teams, Not Managers/Leaders

Early on, some of the Agile frameworks did not address how leaders and/ or managers fit into the process. This is unfortunate, as it resulted in the notion that we either don't need them or they do not provide value in this new environment. As we define new ways of working, it does mean rethinking how we work together, how we create a culture of psychological safety, and how we create a learning culture. Leadership is fundamental to enable the successful delivery of products and services. Build leaders at all levels of the organization. In L. David Marquet's book *Turn the Ship Around! A True Story of Turning Followers into Leaders*, he explains that competence and organizational clarity are the pillars needed for pushing the decision-making authority at the right levels within the organization.[4] This change in mindset is driven and modeled by leadership.

To read more about the importance of the leader's role in Industrial DevOps environments, read Chapter 12: Apply a Growth Mindset.

MISCONCEPTION: Agile/DevOps Requires Deploying Operations Continuously

Yes, we want to release early and often. The goal is to deliver value in the shortest sustainable lead time. Companies like Amazon and Google release thousands of times per day and deploy into operations continuously. That works well for their environment, for their product, and for their customers. But that is not the case everywhere. Each iteration of work—each backlog item, when complete—should meet all the desired quality standards and the acceptance criteria such that it works as intended. This means it could be released if you have products or users where this is feasible and desired.

However, in cyber-physical systems, while you work with this mindset, you might be releasing into a large system-of-systems environment to be fully integrated. The goal is to identify what we can learn earlier. Maybe the hardware isn't ready, but there is a simulated environment you can learn in. We cannot release it until the full system—hardware and software—has been fully integrated, meets safety requirements, and successfully passes the full launch checklist.

To learn more, read Chapter 8: Iterate, Manage Queues, Create Flow and Chapter 10: Integrate Early and Often.

MISCONCEPTION: Agile/DevOps Requires Everyone To Be Colocated

According to Principle 6 of the Agile Manifesto, "The most efficient and effective method of conveying information to and within a development team is face-to-face conversation."[5] Research emphasizes the value of face-to-face conversations. A great deal of communication is nonverbal, and building relationships and trust happens more easily when we have shared experiences.

This does not mean that *everyone* has to be colocated or that they have to be colocated all the time. Distributed teams are common now, and as you scale across teams of teams, the workforce will span across geographical locations. We are in a good position today, with the emerging digital technologies and tools, to be more effective when distributed. There are many more tools at hand to make distributed and virtual teams work.

Be prepared and be intentional with communication and building relationships when teams are not colocated. According to McKinsey research,

> Without the seamless access to colleagues afforded by frequent, in-person team events, meals, and coffee chats, it can be harder to sustain the kind of camaraderie, community, and trust that comes more easily to co-located teams. It also takes more purposeful effort to create a unified one-team experience, encourage bonding among existing team members, or onboard new ones, or even to track and develop the very spontaneous ideas and innovation that makes agile so powerful to begin with.[6]

To learn more about organizing teams to deliver value, read Chapter 4: Organize for the Flow of Value.

MISCONCEPTION: Agile/DevOps Practices Are Only for Software

Agile and DevOps may have started in software, and Lean may have started in manufacturing. But just because these practices started there does not imply that they are not applicable in other functions. Agile and DevOps started in software as developers were looking for better ways to manage their work and be responsive to changing priorities and customer needs. Lean began in manufacturing as a mechanism to reduce waste and improve quality. Over the years, these practices have permeated into other functional areas, and with the evolution of digital tools and capabilities, these practices have evolved into hardware and manufacturing. It is important to extend the Industrial DevOps principles and mindset across the value stream to ensure teams are continuously improving flow and the delivery of value, which requires more than software when building cyber-physical systems.

As you've read, Chapter 1: Toward a Practice of Industrial DevOps talks about Agile and DevOps applicability to cyber-physical systems.

MISCONCEPTION: Agile/DevOps Does Not Work with Safety-Critical Systems

Another misconception is that Agile/DevOps does not work with safety-critical systems. This misconception is likely related to other misconceptions, such as quality and planning. Agile/DevOps development efforts have improved quality, and planning still happens. In fact, due to improved built-in quality and system traceability, Agile/DevOps has the opportunity to improve the safety of systems, as noted in an *IEEE Software* article:

> High dependability software systems must be developed and maintained using rigorous safety-assurance practices. By leveraging traceability, we can visualize and analyze changes as they occur, mitigate potential hazards, and support greater agility.[7]

Based on experiences and research from the industry, we proposed in our 2022 Industrial DevOps paper to the hardware engineering community that "applying Industrial DevOps principles should be required when building safety-critical cyber-physical solutions to lower risk and improve quality."[8] Furthermore, Agile/DevOps has already been implemented in safety-critical systems and is increasingly scaling and maturing. You will need to ensure your organization has overcome the other misconceptions for this to be effective.

To learn more about testing and building in quality, read Chapter 11: Shift Left.

MISCONCEPTION: Agile/DevOps Requires You to Complete a Whole System in Two Weeks

The capabilities of cyber-physical systems continuously evolve as the product is built. Given that these solutions include software and hardware, it is recognized that a complete system is not fully integrated and tested within a two-week period. We are focused on what we can learn in the time box, and the capabilities of the system are decomposed in such a manner that every two weeks, there is better visibility into what is working, what has been successfully integrated, and, based on observation and data, what next steps need to be taken. This approach provides ongoing learning in short time frames, so when course correction is needed, the impact on the schedule and cost is less than if discovered six or twelve months later, which has often been the case with traditional development approaches.

To learn more, read Chapter 6: Implement Data-Driven Decisions and Chapter 10: Integrate Early and Often.

KEY TAKEAWAYS

- Planning is an essential function in Agile/DevOps. In some cases, teams may be planning more if they have been dependent on managers in the past to hand them their plan.
- Build quality in. Agility and responsiveness depend on quality; otherwise, rework and long schedule delays will be the result.
- Agile does not mean no documentation. Progress is measured on demonstration or results of something learned, not how much documentation has been created. However, systems require documentation that evolves with the system.
- The goal is to learn, not build and deploy entire systems in two weeks.

QUESTIONS FOR YOUR TEAM TO ANSWER

- When have you experienced or witnessed any of these misconceptions? Based on what you have just read, how would you respond to each misconception?
- Does changing the goal from delivering systems every two weeks to demonstrating learning every two weeks change your perspective? How would you explain this mindset to your leadership team?
- How could intentional learning impact your delivery?

COACHING TIPS

- Begin with a plan and use empirical data to further inform the plan. Understand that as a self-managing, self-organizing team, the team

is responsible for planning their work in alignment with the organization's priorities.
- You need as much documentation as is required to design, communicate, and sustain your system. Discuss as a team what level of documentation is needed for your system and why it is needed.
- With your team, discuss which misconceptions you are encountering. Capture an agreed-upon response to that misconception. The next time you encounter that misconception, you will be better prepared to respond.
- Leaders are fundamental to building highly engaged, empowered teams. As a leader, write down the actions you have taken to empower your team to be leaders. How have you provided organizational clarity and the technical competence necessary for your team to succeed?

PART II
THE PRINCIPLES AND PRACTICES OF INDUSTRIAL DEVOPS

CHAPTER 4

ORGANIZE FOR THE FLOW OF VALUE

> **PRINCIPLE 1:** Organizing for flow aligns your multiple product teams for regular demonstrations, feedback, and delivery of value.

The purpose of any business is to deliver value to its customers. One key way to keep those customers is to deliver value *faster* than your competition. Digital natives such as Facebook, Amazon, Apple, Netflix, and Google (often referred to as FAANG) have been some of the most successful businesses at this. Companies like these have selected their organizational structures intentionally, moving away from deep hierarchical structures to flatter structures organized around the flow of value to optimize for products and services delivered at speed to their customers. The digital age is enabling companies to quickly innovate, and the companies that learn the fastest win.

For cyber-physical systems, it is necessary to visualize and understand the steps of the value stream that take us from customer need to delivery of the product. There are a series of steps necessary to develop, manufacture, release, and sustain the system. The challenge is that, historically, many organizations have been organized by functional areas rather than around the value stream, which resulted in many hand-offs from one area to another. This led to extensive documentation to communicate between the functions. To address complexity and improve responsiveness, organizations need to reduce the functional hand-offs to enable teams to drive toward business outcomes.

In this chapter, you will learn how to align multiple product teams to enable the flow and delivery of value.

Organizational Structure

It is important to understand the impact of the organizational structure on the flow of value delivered to customers, users, and stakeholders. There are different approaches organizations can take, with each having a set of pros and cons. There is not a one-size-fits-all solution.

There are multiple options for organizational structures, including *flat divisional, functional hierarchy*, and *matrixed*. Selecting the structure for a particular product must be made intentionally based on the priorities of the business. For example, a business whose priority is *speed* to customers with minimal dependencies might choose a flat divisional structure because it has minimal hand-offs. However, this organizational structure does not provide economies of scale. Historically, large businesses building cyber-physical systems with dependencies between business areas prioritized *economies of scale* and selected a functional hierarchy (Figure 4.1). This approach sacrificed speed, as the hand-offs between functions resulted in long lead times and delays.

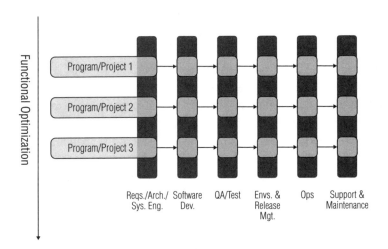

Figure 4.1: Functional Organizational Structure

The current trend in industry has made "speed of delivery" the highest priority, which suggests a flat structure organized around value. However, given the size of many cyber-physical systems, the number of teams required to deliver those systems, and their complexity, we also need a structure that will provide for greater coordination and communication. These needs lead us to a matrix-style organization.

A matrixed organizational structure (see Figure 4.2) aims to support efficiency while maximizing coordination for the business. A matrixed organization has the added benefits of greater emphasis on skill and career development and advancement.[1]

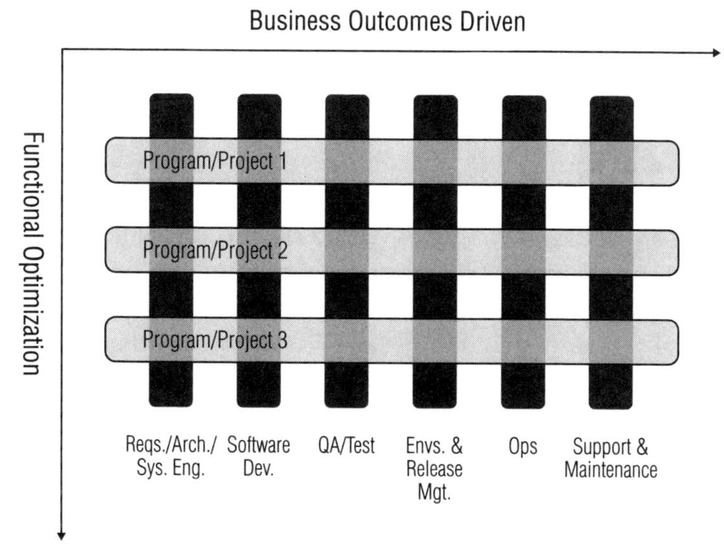

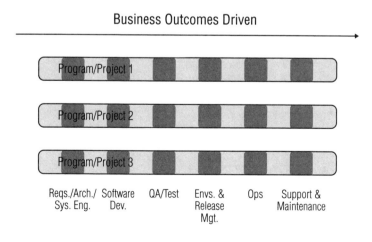

Business Outcomes Driven

Functional Optimization

Program/Project 1

Program/Project 2

Program/Project 3

Reqs./Arch./   Software   QA/Test   Envs. &   Ops   Support &
Sys. Eng.   Dev.   Release   Maintenance
                Mgt.

Figure 4.2: Matrixed Organizational Structure

Alternatively, *product-based* or *divisional* (Figure 4.3) organizational structures can provide the shortest time to market and have a greater customer focus. The divisional structure is organized around the flow of value to customers, where cross-functional teams are designed to deliver products in the shortest lead time while optimizing for rapid learning. With this structure, the organization may lose attention to developing the skills and expertise of the functional homerooms.

Business Outcomes Driven

Program/Project 1

Program/Project 2

Program/Project 3

Reqs./Arch./   Software   QA/Test   Envs. &   Ops   Support &
Sys. Eng.   Dev.   Release   Maintenance
                Mgt.

Figure 4.3: Divisional Organizational Structure

In order to pick the most appropriate organizational structure for optimal flow of value, a business must first map its value stream to understand all the steps in materials and information flow needed to design, develop, test, deploy, and manufacture the products. (We'll talk about value streams later in this chapter.)

As we progress in the digital age, our function-based organizational structures are not supporting speed. After all, functional hierarchical organizations generate monolithic systems. This means one small change can impact the entire system. Now is the time to move away from functional hierarchical organizations and move to flatter structures organized around value.

In theory, it seems like you should be able to just dive in and change your organizational structure; however, if you are working in a large organization, the better approach is to implement incrementally. What we have learned from experience and Conway's law* is that organizations build systems that mirror the organizational structure. For these monolithic organizations, we need to follow what Team Topologies experts Matthew Skelton and Manuel Pais refer to as the inverse Conway maneuver, by incrementally changing the organizational structure to improve the flow within the system and in concert with refactoring the organizational and system architecture into modular components.

Simply moving teams to be organized around value streams will not automatically change the speed of delivery. Those small independent teams will have many dependencies with other small independent teams, bringing delivery to a halt. To successfully handle this problem, we need to evaluate the system, create a boundary around a small portion of the system, refactor the area into modular components, and then pull a portion of the hierarchical organization to create a flat team that can support those modules. Iterate that approach until you have refactored your system and your organization.

Team Composition

The next consideration is team composition. For cyber-physical systems, some of the common practices, such as cross-functional feature teams, can be very challenging. For example, a satellite system can require hundreds of engineers to build. There is both the ground system and the space system. The space system alone has a bus, payload, communications, structures,

---

* Conway's law is the theory that organizations design systems that match their communication structure.

power, thermal, attitude determination, and GNC (guidance, navigation, and control) systems. No single feature team can successfully understand all facets of the system, simply because of its size and complexity.

In the past, organizations tried multiple approaches to resolve the difficulty. For example, we would create component teams that could be decomposed into cross-functional feature teams to support that component. Or we'd create a set of feature teams that supported the payload component.

Obviously, many people trying to scale were experiencing this problem. In 2019, Skelton and Pais published their response to the problem with *Team Topologies: Organizing Business and Technology Teams for Fast Flow*. In the book, they discuss the fact that cognitive load is a real problem and provided some new ideas on organizing business and technology teams for fast flow through organizational design and team interaction.

Team Topologies defines several patterns based on four team types (stream-aligned, enabling, complicated subsystem, and platform) and three modes of team interaction (collaboration, x-as-a-service, and facilitating).[2] We are not going to recreate the whole book here, but we do want to provide some information on the team types and interaction modes in the context of the large satellite system we mentioned earlier (see Tables 4.1 and 4.2).

Table 4.1: Four Team Types of a Satellite System

Team Topologies Team Type	Description	Satellite System Team
Stream-Aligned Team	Aligned to a flow of work from (usually) a segment of the business domain.	Payload team
Enabling Team	Helps a stream-aligned team to overcome obstacles. Also detects missing capabilities.	Cyber security team
Complicated-Subsystem Team	Where significant mathematics/calculation/technical expertise is needed.	Guidance, navigation, and control Team
Platform Team	A grouping of other team types that provide a compelling internal product to accelerate delivery by stream-aligned teams.	Continuous delivery pipeline team

Table 4.2: Three Interaction Modes of a Satellite System

Mode of Interaction	Description	Satellite System Example
Collaboration	Working together for a defined period of time to discover new things (APIs, practices, technologies, etc.).	The payload team collaborates with the guidance, navigation, and control Team to transmit navigation signals over S Band.
X-as-a-Service	One team provides and one team consumes something "as a service."	The guidance, navigation, and control team can provide navigation data as a service to other components.
Facilitation	One team helps and mentors another team.	The thermal team mentors the structures team.

The traditional approach of organizing by functional area created hand-offs and documentation as the primary means of communication. Of course, the teams couldn't hand over their requirement specifications or designs until everything was planned out months and years in advance. When all the systems engineering documentation was done, the documentation was handed over to the software development team to build, and the systems engineering team moved on to the next effort.

Maybe there was a level of efficiency using this approach decades ago, when the cost of change was high or when requirements were more stable, but in today's world, with changing priorities, growing advancements in technologies, and evolving missions, this approach has lost its luster.

The shift from the functional-based organizational structures to product-focused value streams is becoming the norm. Organizations define their value streams from a product perspective, executing via cross-functional teams and short development/learning cycles. Short cycles with frequent feedback loops help ensure the team is building the right product. When people use the phrase "fail fast, fail early," what they mean is that recovery is faster and therefore less costly because it has been only a short duration of a couple of iterations or weeks lost. Organizing around value with cross-functional teams working in short iterations results in higher-quality products.

How to Identify the Flow of Value

The term *value stream* originated from the Lean movement to describe the material and information flow needed to deliver value to customers. There are two types of value streams: *operational*, which represent the flow of business from concept to cash, and *development*, which represent the building and operation of systems needed to enable the operational value stream.

A value stream starts with a customer need and goes through a sequence of engineering and management activities that result in value delivered to the customer. That value could be a new product, a newly released capability, or a service provided. To identify your value stream, you must first understand the flow of work your organization goes through to deliver value to your customer. Start by capturing how customer needs enter your system and all the high-level steps that are taken to deliver that need.

Let's look at the value stream of a CubeSat (a miniature satellite). First, the organization starts by understanding the customer and their need. Our CubeSat mission is to improve weather forecasting accuracy. Our targeted customer is interested in this data to improve their predictive models (their need). The initial goal is to iteratively launch two hundred CubeSats within the first fifteen months and reduce the lead time to twelve months. The CubeSats need to be replenished every twelve months while building enhanced capabilities. The initial launch has a limited set of capabilities, which will be enhanced with each additional launch, including enabling capabilities such as a digital twin for each satellite and AI as part of the satellite network. Due to the number of CubeSats in production and the advanced technologies and growth of the organization, we have eight small teams working on the attitude control systems, which includes the configuration, modeling, and digital twin capability. Our CubeSat value stream is illustrated in Figure 4.4. While the value stream for a CubeSat appears to be sequential in nature, each of the steps should be implemented through a series of short, iterative feedback cycles.

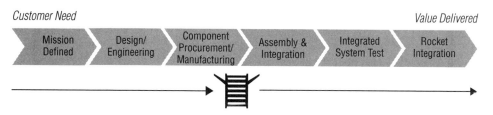

Figure 4.4: CubeSat Value Stream

The overarching value stream needs to be defined to the next-level value streams to better understand which systems are being built to deliver that value. Then the teams can be organized around those systems. Capturing these nested value streams shows alignment from the high-level flow of work of the value stream to the development value streams. Remember: capturing the value stream is not a one-time activity. It is going to change and evolve and is always being improved.

Cyber-physical systems are composed of multiple subsystems, resulting in value streams that are often nested and complex. Therefore, with Industrial DevOps, the value stream is decomposed into smaller *nested value streams* to better understand how different systems are built and then integrated into a whole. Each nested value stream delivers an integrated system as part of the solution.

Figure 4.5 shows how the value stream can be decomposed into nested value streams, with each one able to deliver into an integrated system. For our CubeSat network, we identify nested value streams to reduce dependencies between teams such that each stream can build its features independently yet is able to integrate across the value streams on a regular cadence.

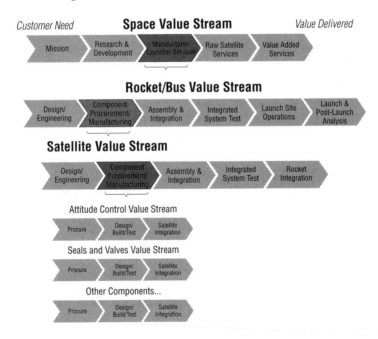

Figure 4.5: Example of Nested Value Streams for CubeSat Constellation

---

* In the original Industrial DevOps papers, these were called value streamlets. The language has since been updated.

Understanding the different nested value streams and how they work together is important for improving communication and the ability to push decision-making to lower parts of the organization—closer to where the work is done.

For each nested value stream, there exists one or more development value streams (Figure 4.6.). A development value stream is an Agile team of teams, including all functions, who collaborate to build the systems within each nested value stream. The development value streams often leverage all four of the Team Topologies team types (as mentioned earlier). An Agile team of teams within a development value stream delivers integrated functionality that is demonstrated regularly and typically has fewer than 150 members (in accordance with Dunbar's number). The software, hardware, systems engineering, and testing are functions within the development value streams and, as defined by the Lean Enterprise Institute, include all the "value-creating and non-value-creating required to bring a product from concept to launch."[3]

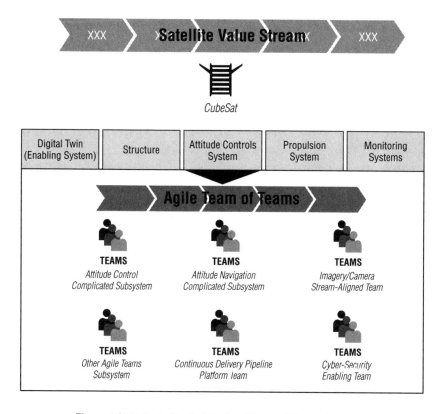

Figure 4.6: Attitude Control System Team-of-Teams Structure

The non-value-created activities are why the practice of continuous improvement is important. These can include defects, overproduction, transport, motion between steps in a process, inventory, or any other task that is adding time to the development of work and is not adding value to the product. Retrospectives from the Agile community and Kaizen events from the Lean community both focus on improving the flow of work across the value stream to improve the speed and value of product delivery.

Each Lean-Agile team has all the skills needed to implement changes from definition to release, specific to their development value stream, to independently deliver user value. During development cycles, nested value streams integrate and test their capabilities often, using a *right-sized* integrated approach. During the early design, integration and testing are done through a series of models and simulated environments. As solutions and options are tested in the simulated virtual environment and there is convergence toward the physical solution, software integration moves into physical hardware.

The value streams may represent different suppliers that are responsible for building different parts of the system and delivering or integrating on a defined and aligned cadence with the rest of the system. For example, if the system is a plane, the primary contractor may be using a supplier to build the wings. Or, if the system is a car, the supplier might be working on the engine. In the case of a satellite, the supplier might be providing the antennae. Whatever the situation, there needs to be defined interfaces and standards for each nested value stream to work independently yet synchronize, integrate, and test on cadence.

The cross-functional team of teams within the value stream integrates and demonstrates their work frequently. When building cyber-physical systems, some teams will be hardware-centric, some will be software-centric, and others may be enabling teams with specialized capabilities, such as integrated logistics support to perform failure modes, effects, and criticality analysis needed by the teams. This Agile team of teams builds integrated features and has all the skills needed to deliver that functionality.

When organizing teams around value, consider how they can work as independently as possible. This means building systems where work can be decoupled so teams can deliver more autonomously, requiring less coordination with other teams. Using well-defined interfaces and architecting for speed are foundational for agility and speed.

In our CubeSat example, the value stream is made up of multiple systems. Each of the systems has integrated capabilities developed by the cross-functional teams (Figure 4.7).

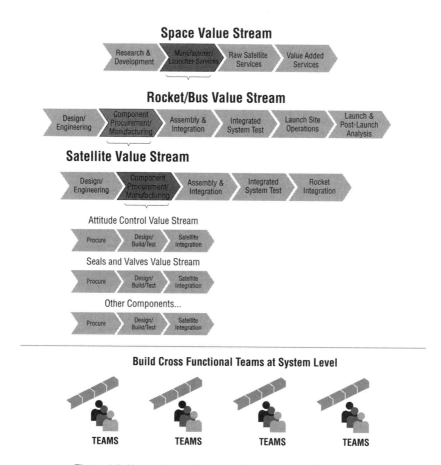

Figure 4.7: Nested Value Stream with Cross-Functional Teams

The cross-functional teams are set up and defined as a stream-aligned team, enabling team, complicated-subsystem team, or platform team. A key consideration for teams performing product development is the impact on manufacturing. The manufacturing teams can provide key insight into optimized materials and assembly options that could dramatically reduce the cost of time in manufacturing the products when we scale. For example, in 2022, we took a class from Joe Justice. He explained that Tesla manufacturing teams are involved in early product design to optimize for efficiency and scalability in production; this led to a design change in their mega-casting technique for producing the rear underbody structure for Model Y, which reduced both production costs and quality issues.[4]

In our scenario, the system features are owned and developed by cross-functional teams as they build the attitude control systems and the imagery functionality of the attitude control system. The platform team will focus on building the digital engineering environment and the setup of integrated tools that the teams use to build functionality. The cyber-security-enabling team ensures the system is adhering to policies and requirements necessary for safety-critical systems.

In our Industrial DevOps environments, the Agile teams doing the work are the ones who make the detailed plan for how the work will be done within their team and will approach that work based on the priorities of the product owner. No one assigns work to the team; the team figures out how to best organize themselves around the prioritized work.

Pollyanna Pixton, author of *The Agile Culture: Leading through Trust and Ownership*, stated in an interview in 2014, "When people own a solution, they feel they are making a valuable contribution to the company where they work. It is no longer just doing what you are told. This is much more motivating and satisfying."[5] It is important that product owners define the priorities of the work that needs to be done; the ownership of how the team will work and coordinate to build those priorities belongs to the team. With that comes learning, ownership, pride, and accountability.

As organizations define their value streams from a product lens, they use this structure to shape and define how they create Agile teams to learn and innovate, provide fast feedback, and easily adjust to changing needs and priorities. Organizing for flow of value must be aligned with architecting for speed and change (Principle 4).

Organize around Value Delivery in Manufacturing

For cyber-physical systems, organizing for value delivery is not just the responsibility of development; it extends into optimizing for manufacturability. Teams on the manufacturing floor are also organized around the flow of value with well-defined and visible processes (Figure 4.8). Their routines can be observed and measured. Metrics are displayed visibly on dashboards and reviewed throughout the day. And, as parts move across subassembly to final assembly, we can see the creation of the physical product.

Flexibility to respond to design changes and limit downtime on the manufacturing line requires agility. Organizing teams around the value stream and flow of work enables them to plan and manage their work as independently as possible to improve the speed of value delivery. Let's look at a couple of case studies that illustrate this idea.

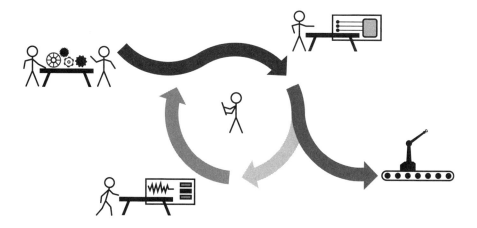

Figure 4.8: Manufacturing Floor

CASE STUDY: Saab Aeronautics[6]

The aerospace defense industry builds some of the largest and most complex cyber-physical systems in the world. These systems are often in the hundreds of millions and sometimes billions of dollars, starting with customer needs, definition to design, integration and testing, manufacturing, and sustainment.

Saab Aeronautics in Sweden needed to deliver fast and keep costs down. They applied Agile principles and practices in the creation of their Gripen E fighter jet to respond to this need. In 2019, their effort included over two thousand individuals applying Agile practices based on Scrum across software and hardware. Their Scrum teams were formed around fighter jet components including fuselage and structures, cockpit, software controls, and engine.

Saab has adopted an Agile approach to organizing itself and its efforts, delivering results more quickly, with higher quality, and at drastically lower cost.[7]

Through the alignment of their Scrum teams and leadership teams, they were able to reduce bureaucracy and encourage decision-making at the lowest possible levels within the organization. Their defined daily escalation path allows teams to share their impediments every day, which rolls up to the highest levels of the organization. At each level, managers address the impediments from the teams. Any issues they are not able to resolve roll up to the next level in the organization until they reach a resolution. Leadership at all levels is actively engaged and committed to supporting the day-to-day efforts of the development teams. This structure has resulted in their ability to deliver an aircraft at low cost with advanced performance and higher quality.

CASE STUDY: BMW[8]

The BMW Group in Leipzig has over five thousand employees who embody modern and digital technologies that result in exemplary capabilities across product development. Based on Mik Kersten's experiences, as presented in his book *Project to Product* and via the blog post "Code on the Road," we have learned of BMW's commitment to integrate and visualize flow across their value stream. They have been able to successfully and "seamlessly integrate production lines with the software life cycle to transform in the digital-first world."[9]

To accomplish this, they first needed to capture the value stream and organize people around it while providing the infrastructure and tools to visualize the workflow. Using continuous improvement and Lean principles as a foundation, they were able to demonstrate how digital transformation ties directly to value creation—not digital transformation for the sake of digital transformation. Instead, they use these enablers to drive toward real business outcomes, and they are seeing results.

Several years ago BMW Group was focused on gaining a foothold in the electric car market and improving market adoption. To support this effort, they desired to better understand the market and how they could use that learning to build out their plan. Building out their digital capabilities, one of the first steps they took was creating "a production architecture to support learning from the market before investing further in automation of the lines. The profitability and product/market fit drove the architecture of the value stream, not vice versa."[10]

It is important that your value stream is aligned to measure your business outcomes, just as the BMW Group organized around their value stream. Using their digital technologies to visualize flow, they adopted a product-centric mindset while targeting specific business outcomes. This Lean-Agile mindset coupled with digital capabilities enabled them to quickly transform how they build cars and create a foothold in the "i3 and i8 production lines without ever having mass-produced electric cars or carbon-fiber bodies before."[11]

BMW wanted to understand their value stream networks and gain insight into the flow of production. This would help them see the bottlenecks across their value stream network and know where they needed to improve the system. Applying Lean principles, they focused on removing the bottleneck to improve flow and improve the lead time to deliver value faster.[12]

GETTING STARTED

- Assess your current organizational structure. Identify the pros and cons of that structure. Identify how the structure might be improved to improve collaboration and flow.
- Identify your value stream(s) and the systems built within those value streams.
- Using the different team types, define your Agile team-of-teams structure that develops the product and components within the value stream.
- Define how you will measure success—that is, your business outcomes.

KEY TAKEAWAYS

- Begin by organizing around value. Along with having a product-based mindset, ensure your teams are organized such that they have reduced dependencies and have as much autonomy as possible.
- Identify your stakeholders needed to execute value stream workshops to determine the flow of value.
- Organizational structure is important to ensure we are organizing around value delivery and to enable decision-making at the right levels, along with improving the flow of information and collaboration.

QUESTIONS FOR YOUR TEAM TO ANSWER

- What are the business outcomes your organization would like to achieve?
- What is your current value stream?
- How might you organize your Agile teams around the value stream?
- How will you leverage the Team Topologies team types? Why does this matter?
- What are some criteria to consider as you begin to organize the teams?
- What problem is this helping you solve?

COACHING TIPS

- Just because you have shifted from a project-to product-based mindset does not necessarily mean your teams have been organized around the value stream. Start by understanding your value stream and the solutions needed to enable it, then organize people around those solutions.
- For more reading, refer to *Team Topologies: Organizing Business and Technology Teams for Fast Flow* for detailed descriptions of four fundamental teaming structures. This book addresses software and IT team structures. Use what you have learned from this chapter to include hardware and embedded software team members in your team structure.

CHAPTER 5

APPLY MULTIPLE HORIZONS OF PLANNING

> **PRINCIPLE 2:** Apply multiple horizons of planning to address scaling and complexity while leveraging ongoing experimentation and learning.

Five years ago, when we were first developing the idea of Industrial DevOps, we would have never predicted COVID-19 or the impact it would have on the world. In March of 2020, our employers sent people home to work virtually for what we assumed would be a week or two. As of March of 2023, many people were still working virtually. Some companies went out of business, and other companies thrived.

Ultimately, the COVID-19 pandemic is a good example of why predictive planning doesn't work: nobody has a crystal ball. However, *no* plan is not helpful either. We have been working on this book for over a year, and you would not be reading this now if we did not have a plan.

The answer to using planning effectively so that it does not become a hindrance is to move from predictive planning to empirical planning, an approach that involves using data and feedback to inform decision-making and adjust plans as needed. In addition, when developing cyber-physical systems, we need to have *multiple* planning horizons, not just one. Most cyber-physical systems take months to years to build. Thus, a single plan is unlikely to be effective.

Applying multiple horizons of planning is similar to how communities are built. While parts of the community will grow organically over time, there is an infrastructure in place for each home to build from and a plan or general framework as the community emerges. Homes are not delivered all at once but incrementally. And each home, while built within the infrastructure of the community boundaries, also has its own unique features that meet the specific desires of the home buyer. The building of this community requires multiple levels of planning with different levels of fidelity at the implementation level.

In this chapter, you will learn about the different levels of planning that are designed to address scaling and complexity, with examples as to how this might align with the product backlog of cyber-physical systems. We will begin with how the planning mindset has shifted, understanding the past as we build the future.

Predictive vs. Empirical Process Control

Two main process control models for delivering products and services are predictive process control and empirical process control. Predictive process control models predict the future based on historical information and the expectation of little change, along with stage gates to determine state and manage the flow of work. Empirical process control models use current observations to determine state, adapt to change, and manage the flow of work. When the process is simple and repeatable, a predictive process control model can be used; however, for complex processes with greater unknowns, an empirical process model is recommended. Figure 5.1 shows the spectrum of process control.

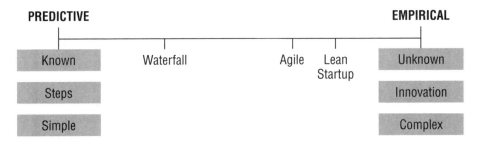

Figure 5.1: Predictive vs. Empirical Process Control

The waterfall approach that so many cyber-physical systems are built with is based on a predictive process control model. This form of planning defines a fixed schedule for execution at the start of a project and has been utilized across multiple industries for decades to manage both simple and complex projects (e.g., the Hoover Dam).

The key issue with this approach is the assumption that we can predict the implementation schedule at the start of the project, which is when we know the least about the solution we are building. The areas of the cyber-physical system that are software-centric are especially difficult to predict because teams are designing the solution as it is implemented and there is an ongoing learning and improvement feedback cycle that happens during development. And, with hardware design and development being done in models and simulated environments, much of the early hardware development is digital and easier to change.

Dr. Winston Royce, an expert in the space domain and one of the originators of the waterfall approach, worked on various efforts involving the development of complex, large-scale software packages for mission planning, commanding, and post-flight analysis. Dr. Royce witnessed varying degrees of success with the predictive process control model in complex product development, which led him to publish the article "Managing the Development of Large Software Systems," where he described the problems associated with the model and the need for an empirical process control with regular feedback.[1] Unfortunately, the industry misunderstood the intent of the article and went on to use the waterfall life cycle as the de facto standard for the development of many complex systems development efforts for over fifty years.

The waterfall life cycle utilizes a series of stage gates at the close of each phase, which include requirements, design, implementation, verification, and maintenance (Figure 5.2). Depending on the context of the product and domain, there may be more or fewer phase gates. The DoD utilizes a phase-gate approach to acquire and build systems. Success of the product is defined by the completion of each phase gate.

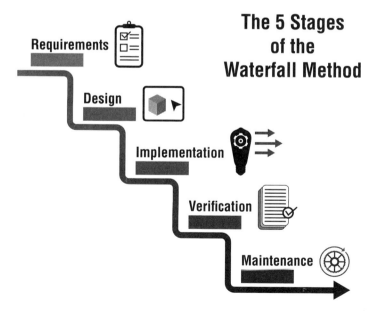

The 5 Stages of the Waterfall Method

Requirements

Design

Implementation

Verification

Maintenance

Figure 5.2: Royce and the Steps for Solutioning a System

The problem is that as systems grow in complexity, these phase gates can last years and the predictive approach breaks down. As Dr. Royce's paper outlined in 1970, the waterfall approach for complex systems is risky and invites failure.[2] While this predictive form of planning may have worked in the past for some efforts—which we question—with the advent of new technologies and the desire for greater innovation and more frequent delivery of capabilities, this is no longer the case.

We know change is going to happen; therefore, while we still plan, how we plan must be different from what many of us may have experienced in the past. Only in environments where there is no change and where requirements are well understood will detailed project schedules with ten thousand or more lines of tasks work. This level of inflexibility may no longer be in your best interest but may be in the best interest of your competitor.

We know that no one can predict the future, and in cyber-physical systems, we need room to adapt to change as we learn.

Multiple Horizons of Planning

Applying multiple horizons of planning is an empirical planning approach in which we have multiple plans at varying time horizons and with different levels of granularity. The key difference empirical planning has from predictive planning is the inclusion of a regular business rhythm to review the empirical data with the intent to use the data to update the plan and the level of granularity when we plan.

Historically we would have had a detailed project plan that extended for many years. When we evaluated the resulting outcome, the goal was to change how we are working to meet the plan instead of changing the plan to meet the way we are working. We have learned the importance of responding to changing mission needs, customer needs, new technologies, and changing environments—that is, "responding to change over following a plan."

Successful planning means understanding where you are headed with a high-level road map, with the details emerging as those carrying out the plan implement and learn (Figure 5.3). This allows teams to identify long-lead hardware needs while still ensuring flexibility and responding to change at the lowest level of implementation. Each horizon yields empirical data, demonstrating progress of the product and identifying necessary adjustments.

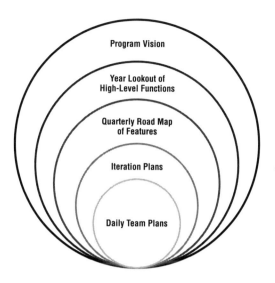

Figure 5.3: Multiple Horizons of Planning

If the picture in your head is a traditional predictive schedule that contains multiple years, then you have the wrong picture. A better picture is illustrated in Figure 5.4, which describes NASA's road map to human space exploration. With each milestone, they reevaluate the plan and update based on what they have learned.

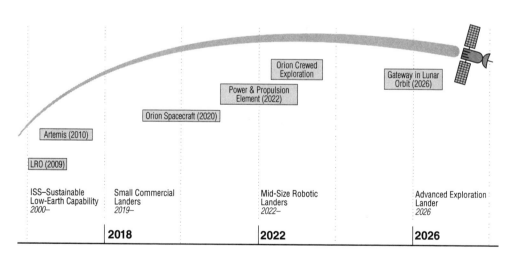

Figure 5.4: NASA's Road Map for Human Space Exploration

Source: Adapted from Foust, "NASA Roadmap Report Provides Few New Details on Human Exploration Plans," SpaceNews. September 25, 2018. https://spacenews.com/nasa-roadmap-report-provides-few-new-details-on-human-exploration-plans/

This approach has been used by the military for decades. During the Gulf War in 1991 when US forces witnessed Iraqi forces crossing the Saudi border into Khafji, they were able to quickly assess the situation, develop a response strategy, and execute that strategy. After they observed the forces, they oriented themselves to stop the advance, decided on a course of action, and launched a successful counterattack.[3]

The military refers to this model as the observe-orient-decide-act (OODA) loop, and they use it frequently to respond to changing and unpredictable mission situations. They go in with a plan based on the information they have at the time, realizing that when boots are on the ground, those on the front lines will quickly adjust and pivot as new information is available.[4] Business environments today are quickly learning this model for their own survival. With this style of empirical planning, the planning horizons extend across the entire effort, which could range from months to years. For very large safety-critical cyber-physical systems, the timeline is often a year or more.

The *product vision horizon* provides a line of sight to business objectives for all stakeholders in the organization and will answer the questions "What are we building? Why are we building? Who are we building for?" and include a range for when we plan to deliver with some defined milestones or targets for delivery.

After we define the product vision, we need to define an overarching product plan for the life of the system. This could be months to years to, in some cases, decades if you consider the life of some nuclear power plants. The longer the planning horizon, the lower the fidelity in the plan. You may be thinking that this sounds a lot like a waterfall, multiyear predictive plan. The key difference is the level of fidelity. Industrial DevOps advocates for high-level milestones that align to vision versus high-fidelity, detailed activities. Once we determine the length of system development, then we can determine the planning horizon to start with. Large systems such as rockets typically take many years and have multiple dependencies along the supply chain, requiring a lengthy starting planning horizon. For a relatively small system that can be completed within a year, we would begin at the annual level.

If we assume a small system, we begin with *annual planning*, the year-out look of high-level functionality. This horizon focuses on the major capabilities we plan to design, develop, test, and stage this year. During this level of planning, we are asking, "Where do we want to be at the end of the year?" This is not a detailed plan to the task level. It focuses on the prioritized functionality that is expected. We anticipate change and reprioritization based on change, so we need to plan at a level granular enough to know the direction we are heading yet light enough that responsiveness is still achievable.

To offer more flexibility and have shorter learning cycles, we break the planning into *quarterly cycles* (or shorter cycles based on the business needs) and decompose the year into a quarterly road map of features. This is significant because this is where we decompose our big rocks into smaller efforts and focus on what we need to learn each quarter to reach our bigger goals. The length of this repeating cycle should be reflective of the needs of the organization. Quarterly road maps with a forecast of prioritized capabilities begin to take shape, making the path visible during this horizon of planning. This cadence is important in the planning of cyber-physical systems, as it provides the opportunity to ensure alignment across the team of teams and to identify integration points between hardware, software, and suppliers.

The next horizon of planning is aimed at enabling small cross-functional teams to focus on a set of activities and goals. This is the heart of where the work is performed. The typical time horizon for this cycle is between one to four weeks, with a preference toward smaller timelines. This planning horizon is often referred to as a *sprint* or *iteration*. Most organizations are familiar with the sprint or iteration by now, which was made popular by the Scrum framework. The iteration plan contains small pieces of work, often framed into a construct known as user stories, which are demonstrated when complete for user feedback. This level of detailed planning occurs with full visibility into the activities needed to be completed across the team to meet the goals of the iteration. There is a high fidelity of all activities and timelines to complete those activities.

Finally, the smallest planning horizon is *daily planning*. Each day, teams share with each other what tasks they completed the day before, what tasks they plan to complete today, and what impediments may keep them from completing those tasks. The intent is not a status report but rather how the team is collaborating to achieve the iteration goal. Since the work has shared ownership by the team, daily collaboration is important to achieve the iteration goals. If you find that your teams are not getting any value out of the daily planning session, known as a daily stand-up, evaluate how the work is organized. If every member of the team has an independent story, as opposed to the entire team working on a single story (where they swarm on tasks), then it is likely the members of the team are working in parallel instead of collaboratively, which is less effective.

As you decompose both time and the system elements, we recommend identifying opportunities to complete minimum viable products (MVPs).

There are multiple interpretations of what an MVP is. For cyber-physical systems, we use the definition that an MVP should have enough features that the team can validate their hypothesis. It is not necessary for the MVP to be able to be deployed. An MVP for our CubeSat example could be that we want to validate that we can extract pitch and roll data. To do so, we connect to a Raspberry Pi and use it as a mini flight computer. Next, we write a Python script and validate that we can indeed extract the information from our makeshift flight computer. The largest barrier to teams implementing MVPs is lack of knowledge about how to decompose the system and implement small pieces of functionality.

Decouple Time and Scope

Traditional planning tightly integrates time and scope, which results in a lot less flexibility in planning the development of products and services. Separating these attributes allows us to move scope around more easily, assuming we do not have a specific dependency. Table 5.1 describes what we mean by time and scope.

Table 5.1: Time and Scope Defined

	Time	Scope
1.	Program/product plan (entire time box)	Product vision
2.	Multiyear plan (1–5-year time box)	Epic (business outcome)
3.	Annual plan (1-year time box)	Epic (business outcome)
4.	Quarterly plan (12–13 weeks)	Feature (business outcome that fits within quarter)
5.	Iteration plan (1–4 weeks)	User story (user outcome that fits in iteration)
6.	Daily plan (8 hours)	Task (individual outcome for today)

To decouple time and scope, begin by decomposing scope independently of time, identifying dependencies, and then planning scope into a planning time box. A good approach to achieve this is through the development of behaviors and scenarios that we can test:

1. Identify subsystem and components.
2. Define desired behavior of subsystem and components.
3. Define scenario using the *given*, *when*, *then* format.
4. Define acceptance criteria.

If we apply this to our CubeSat example:

1. **Identify subsystem and components:** The attitude control system (sensors, actuators, and communication interface).
2. **Identify desired behavior:** The attitude control system's desired behavior is to control the orientation of the satellite in space.
3. **Define scenario:** *Given* the satellite is in orbit around the Earth, *when* external disturbances or perturbations are encountered, *then* the attitude control system should promptly and autonomously adjust the satellite's orientation to maintain the desired attitude.
4. **Define acceptance criteria:**
   a. System has access to sensor data.
   b. System calculates attitude data based on sensor measurement.
   c. System runs control algorithm to determine computed error.
   d. System commands actuators using computed error.
   e. System actuators adjust to specified attitude.

As we've stated, failure to decompose the scope of work is one of the top reasons cyber-physical teams struggle with implementing Agile/DevOps principles and practices. However, there are a number of proven patterns for decomposition of cyber-physical systems, as illustrated in Table 5.2.

Table 5.2: Patterns for Decomposition

	Pattern	Scope
1.	Work flow steps	Break out all of the steps of the work flow required to deliver value
2.	Business rule variations	Accomplishment of different business rules

Table 5.2 (continued)

Pattern	Scope
3. Major effort	Large-effort items can often be split, where the first one is the instantiation of capability and the remaining continue to improve
4. Simple/complex	Capture simplest version of feature and complete remaining to add complexity
5. Variations in data	Data variations, such as data sources, complexity, language variants
6. Data methods	Split by the user interface itself
7. Deferring system qualities	Begin with a simple capability and add the system qualities incrementally
8. Operations	Order of operations, such as CRUD (create, read, update, delete)
9. Use case scenarios	Split by goals or scenarios

While there are multiple ways to decompose the system, there is one way that we would discourage: the Big Bang functional hand-off, as illustrated in Figure 5.5.

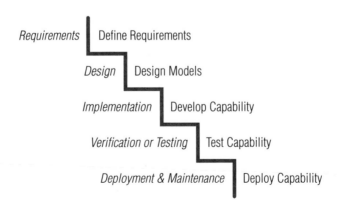

Figure 5.5: Big Bang Solution

Product Vision

You'll notice that in our illustration of multiple horizons of planning (look back to Figure 5.3), there is one all-encompassing planning horizon: the product vision. It is essential to start any planning by aligning with your company's vision and mission. Know your North Star. People want connectedness to purpose and to know that they are part of something significant. That is the advantage and purpose of working in the cyber-physical world. Many of these systems—fighter jets, autonomous vehicles, space systems, sensors, radars—include requirements that help ensure the safety of society or, in some cases, provide features that make life easier or better for customers (like being able to watch your home while traveling). These systems and their purpose drive a sense of connectedness between the work performed by the people and the mission of the company.

Examples of vision statements of well-known companies are below:

- **SpaceX:** "You want to wake up in the morning and think the future is going to be great—and that's what being a spacefaring civilization is all about. It's about believing in the future and thinking that the future will be better than the past. And I can't think of anything more exciting than going out there and being among the stars."[5]
- **Bosch:** "Invented for life: we want our products to spark enthusiasm, improve quality of life, and help conserve natural resources."[6]
- **Chevron:** "Our purpose is to develop the affordable, reliable, ever-cleaner energy that enables human progress."[7]
- **Apple:** "Bringing the best user experience to its customers through innovative hardware, software, and services."[8]
- **NASA:** "NASA explores the unknown in air and space, innovates for the benefit of humanity, and inspires the world through discovery."[9]
- **Amazon:** "Amazon strives to be Earth's most customer-centric company, Earth's best employer, and Earth's safest place to work."[10]

Being part of a team who builds systems with a mission that impacts society can be highly motivational. Leadership must clearly articulate the mission to ensure all those involved understand how their efforts contribute to the success of this mission. Vision and clarity are necessary for building resilient organizations. Plans will change. The vision stays constant, like the North Star, and "the vision is part of the team's mission that is non-negotiable."[11]

When working with organizations seeking to be more Lean, Agile, and adaptive, we start with the "why" and take time to understand the strategic road map. What are the near-term gains we need, and where are they headed in the next three to five years? The strategy is around building systems that deliver value to the customer and building those systems faster, at the speed of need.

The Lean UX Canvas is an excellent tool to capture business needs, a solution summary, and benefits. The Lean UX Canvas in Figure 5.6 provides highlights of our CubeSat problem statement and our approach to meeting our customers' business needs. You can see our hypothetical business problem is the need to increase production rates of our CubeSats. The business is iteratively launching two hundred every fifteen months. Part of the challenge is that our CubeSats last only twelve months. Increasing the rate of production with shorter lead times will enable the company to launch closer to the end of life of the existing satellites that have been launched and meet the customers' growing needs for more data to improve their analytics and forecasting to their user community.

Business Problem/Mission Need	Solution Ideas	Business Outcomes
What business have you indentified that needs help?	List product, feature, or enhancement ideas that help your target audience achieve the benefits they're seeking	What changes in customer behavior will indicate you have solved a real problem in a way that adds value to your customers?
There is a growing demand for improved weather data with reduced costs and increased production rate.	Start in a smaller niche area—impove weather forecasting accuracy and data analytics. Iteratively launch 200 CubeSats over 15 months and improve operations to reduce lead time to 12 months or lower; Low Earth Orbit solution.	Starting with our CubeSat weather mission we will build from this experience as we scale with reduced costs and faster time to market.

Users & Outcomes		User Benefits
What types of users and customers should you focus on first?		What are the goals our users are trying to achieve? What is motivating them to seek out your solution? (e.g., do better at my job OR get a promotion)
Weather forecast customers who want better, more accurate data to improve decision-making and reporting.		Provide timely weather imagery data and broaden the coverage so that they can provide more accurate reports to their user community.

Hypotheses	What's the most important thing we need to learn first?	What's the least amount of work we need to do to learn the next most important thing?
Combine the assumptions from 2, 3, 4 & 5 into the following template hypothesis statement: "We believe that [business outcome] will be achieved if [user] attains [benefit] with [feature]." Each hypothesis should focus on one feature. We believe that we will be able to scale our business if our weather forecast customers provide more accurate reports to their user community by receiving more timely data with wider coverage.	For each hypothesis, identify the riskiest assumption. This is the assumption that will cause the entire idea to fail if it is wrong. Performance constraints of the current design; limitations of imagery/resolution.	Brainstorm the types of experiments you can run to learn whether your riskiest assumption is true or false. Use of models for early-stage design study and simulation of the CubeSat mission.

Figure 5.6: Lean UX Canvas Example: CubeSat

Source: Lean UX Canvas template is used with permission. Jeff Gothelf and Josh Seiden, *Lean UX: Designing Great Products with Agile Teams*, 3rd Edition (O'Reilly Media, 2021).

Multiyear Lookout of High-Level Functions

Now that our vision is clear and we have our North Star, what do we do? In some of the purist Agile mindsets, the thought is that teams should directly begin developing in iterations. However, this doesn't make sense for many large, complex cyber-physical systems, which can take years to build. For example, the original Orion spacecraft contract was awarded to Lockheed Martin in 2006 but wasn't set to be complete until 2020, fourteen years later. That required teams to build a road map and a high-level program plan (or multiyear lookout) that extended throughout the life of the contract.

SpaceX is considered a digital native that embraces Agile mindsets at all levels of the organization. They planned and began working on Starlink, a satellite constellation, in 2015. Eight years later, Starlink is still growing and evolving, and they are using more than an iteration plan to close on goals and objectives.[12]

In general, those building cyber-physical systems need an end-to-end plan whose fidelity reduces the further it extends. The road map and program plan contain high-level desired functionality of capabilities that need to be decomposed over time. Because cyber-physical systems have significant hardware, the road map may include long lead items, milestones, and a critical path (the sequence of stages that determines the minimum time needed for an operation). See a sample multiyear road map in Figure 5.7.

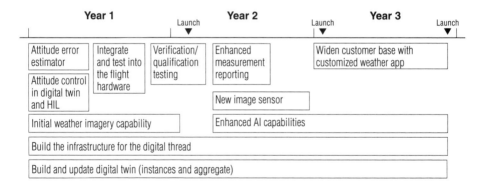

Figure 5.7: Sample Road Map for CubeSat Mission

Annual Road Map of Features

This takes us to the next level of planning, which is annual planning (i.e., what can be accomplished in the upcoming year). The annual plan and road map begin by reviewing the program plan and identifying which big rocks to decompose into elements of value that can be completed within the year. Just as with the program plan (or multiyear road map), the further out the effort is planned, the lower the fidelity of the effort.

Figure 5.8 shows that our annual capabilities road map has been further decomposed into a quarterly road map of features. Near-term planned efforts have the highest degree of fidelity. The following is a notional example for our CubeSat product. Teams building large cyber-physical systems may have a hundred or more features each quarter. Please keep in mind, these are examples and context matters. We may not need quarterly rhythm. Planet Labs has launched 462 CubeSats since 2013 and launches new ones every few months. They don't need a quarterly plan to build their CubeSat, but they may have one to look ahead and evolve their mission.

Quarter 1	Quarter 2	Quarter 3	Quarter 4
Feature: Develop attitude error estimator and integrate into the processor-in-the-loop environment	**Feature:** Incorporate attitude control algorithms in partial digital twin & hardware-in-the-loop	**Feature:** Develop image-processing improvements and test using the camera PIL environment	**Feature:** Integrate and test image-processing updates into the flight hardware
Feature: Build the infrastructure for digital thread			

Figure 5.8: Annual Road Map Broken into Quarters for CubeSat

As teams further refine features each quarter and get closer to implementation, acceptance criteria are added for each feature that defines how that feature will be demonstrated and what requirements must be met for that specific work item (Table 5.3).

Table 5.3 Feature Example with Acceptance Criteria

Feature	Acceptance Criteria
Determine attitude error estimator and integrate into processor-in-the-loop environment.	Incorporate into the target software environment; demonstrate on a processor-in-the-loop environment/simulation (e.g., hybrid cyber-physical twin, emulated processor/other subsystems). Demonstrate the feed to the attitude controller.

The annual plan provided the goals and objectives for the year further decomposed into quarters, but it is not fine-grained enough to execute. This takes us to the next horizon of planning, the quarterly plan. The quarterly plan provides visibility into what the organization can execute on for the next quarter and has higher levels of fidelity of work that can be shared with teams who are slated to execute the work. It uses a technique like rolling-wave planning (Figure 5.9).

Rolling-wave planning is a project management technique that involves planning and executing a project in phases or waves, with detailed planning and execution for the immediate future and broader planning for later stages. Many businesses perform rolling-wave planning at quarterly boundaries because it aligns with a typical financial management pattern.

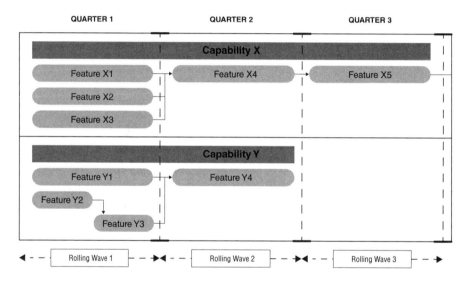

Figure 5.9: Rolling-Wave Planning Example

Quarterly Plan

The quarterly planning horizon provides a useful framework to focus on and get clarity on setting and achieving organizational goals. It also helps teams and individuals to stay focused, adaptable, and accountable. Finally, quarterly plans provide critical building blocks for creating our longer-term plan.

In the following example, the team working on the space ground communications has started modeling a use case that includes the attitude management of the CubeSat (Figure 5.10). Modeling the system is part of their backlog work and evolves with the technical work. The team will zero in on the development of the attitude control system and begin shaping the work to build out that part of the system.

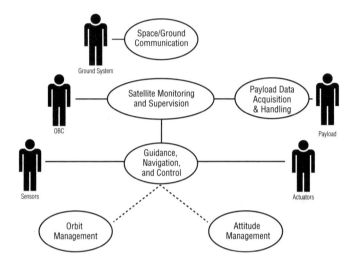

Figure 5.10: CubeSat Space Ground Communication Use Case

The CubeSat team's quarterly road map, illustrated in Figure 5.11, is constructed at a level high enough to provide sufficient detail for stakeholders to have an idea of the path forward while still allowing adaptability and reprioritization. The feature-level work is decomposed into smaller backlog items that teams will work on.

A key goal of the quarterly plan is to visualize work across teams with enough detail to address dependencies between them. The intention at this level of planning is to reduce and remove these dependencies so teams can work as autonomously as possible to improve flow and the delivery of functionality. Before implementation of work can begin, the backlog items have a definition of ready that must be met. The definition of *ready* is a set of criteria that helps the team understand the work item. Common practice includes:

- A description of the item.
- Acceptance criteria.
- Estimated in a relative size (e.g., story points).
- A behavior-driven test utilizing *given, when, then*.

Quarter 1 Sample

Feature	Iteration 1	Iteration 2	Iteration 3	Iteration 4	Iteration 5	Iteration 6
Develop attitude error estimator and integrate into the processor-in-the-loop environment.	Verify that attitude sensor measures current state and feeds data to attitude navigation.	Generate state estimates based upon attitude sensor measurements in SIL.	Determine attitude error using state estimate and desired state from navigation in SIL.	Perform characterization of attitude error estimation in the SIL environment.	Port attitude estimation software into the PIL environment.	Perform characterization of attitude error estimation in the PIL environment and validate alignment with the SIL environment.
Build the infrastructure for the digital thread.	Define the architecture of the digital thread (MVP).	Set up of modeling tool and interface with backlog; communicate process.	Build out connection to requirements.	Add change management process.	Create bill of materials (BOM) structure.	Ensure interfaces and traceability.

Figure 5.11: Quarterly Road Map: CubeSat Example

As the plan is executed, learning the terrain begins, and quarterly plans begin to adjust to take advantage of the learning.

In cyber-physical systems, especially early in the development phase of hardware, the flow of new-feature development between system components often needs to be coordinated between and by the teams, who must engage with other teams to understand the integration points between their work and the work of others and dependencies where they exist. This outlook also means working with suppliers to understand at what point components will be delivered and integrated and tested with the rest of the system.

Iteration Level

The general standard for an iteration length is a fixed length of one to four weeks with a preference for shorter lengths. Begin with one length and iterate from there. The iteration level is where the work is executed and the synergy happens as everyone, collaborating with their team, becomes actively engaged in the planning and envisioning process. Teams work with their product owner to refine and prioritize stories in the iteration. It is in this phase where team autonomy happens.

As they shape iteration plans, each team thinks through the steps that need to happen to *develop*, *integrate*, and *test* new functionality. This is very much in line with Agile software development teams and how they have been planning for years. As we scale, we now have significant dependencies to work through with suppliers and multiple subsystems that need coordination and defined synchronization points.

To address hardware development, the team identifies how they will build and use prototypes, digital twins, simulators, and emulators as part of their development. They identify the earlier points of integration between software and hardware development. They identify the major milestone deliverables on the horizon—such as compliance audits, final assembly, test targets, launch dates, flight tests, and supplier deliveries—to ensure their plans align with those defined milestones and to raise risks against targets.

One area that teams struggle with is how to decompose their work into small enough elements that they can fit into the timeboxes outlined above. This problem is especially true at the feature level, where a feature is defined as a capability that the teams can complete within a quarter. While there are multiple patterns of decomposition, one of the most popular is to create use cases at the feature level, then decompose that feature into user stories. A use case typically contains multiple actors and multiple paths through the system. But a user story looks at one actor for one of the paths through the system.

The use case in Figure 5.10 illustrates a feature that supports Space-Ground Communication by implementing an attitude error estimator. The feature can be decomposed into a user story that states as a Ground System (actor), I want to message Guidance, Navigation, and Control (system) to change the attitude (subsystem) so that I can adjust the CubeSat position (activity) to get improved weather data (business benefit).

With each story, teams include acceptance criteria to define the scope. This is what teams demonstrate once they have finished building and testing the story. Without acceptance criteria, there is a lack of clarity on what the feature or story is expected to deliver, and this is when scope creep happens. Without acceptance criteria, the team struggles to understand what "done" looks like.

As teams map out their plan, they have identified their first feature. This feature (see Table 5.4) will determine the attitude error. The team has started identifying stories for that feature with acceptance criteria they will build and test over the upcoming iterations.

Hardware teams understand the importance of build, test, and feedback cycles. They have been doing similar practices for years. With cyber-physical systems, they must pull that success pattern and couple it with the enabling tools, such as models, simulators, emulators, and additive manufacturing. The rise of digital engineering and additive manufacturing practices has enabled for hardware engineering to develop in short cycles and offer a way to give meaningful demonstrations where stakeholders can provide feedback.

Table 5.4: Example of an Iteration Backlog Item for Attitude Controller

Feature	Iteration 1
Determine attitude error so that it may be fed to the attitude controller	**Estimate:** <relative size> (e.g., 5 story points) **Story (1)**: As an Attitude Sensor, I want to measure the current state of attitude so that I can adjust the attitude. **Acceptance Criteria:** 1. Identify error source inputs for the attitude estimator. 2. For each attitude error source, identify the software elements that will need to be modified to provide error estimates to the controller. 3. Update simulation elements for contributors to adjust attitude error. 4. Model changes in software-in-the-loop (SIL) and hardware-in-the-loop (HIL). 5. Perform Monte Carlo assessment of attitude adjustment error estimator in SIL and HIL.

Daily Plan

Daily plans are the final level of planning. These allow teams to collaborate effectively and provide situational awareness of their work. The daily plan focuses on the tasks to be worked on that day by each team member. The team focuses on their highest-priority work and implements solutions daily.

At the close of each day, the team can see empirical data on how they executed against their plan. This telemetry (learning) may demonstrate the need to adjust the plan. The team communicates about impediments, which are worked off either by the team or the next level of leadership.

Teams can also use this empirical data to further inform their daily planning. For example, if a team planned to complete ten tasks today and completed only seven, they may be having an off day. However, if we trend over multiple days in the sprint/iteration to complete ten tasks and completing only seven, we can apply that knowledge to adjust future planning.

The same learning and empirical data occur at the sprint (iteration) level. If teams complete only 70% of what they planned, they should use that data to adjust the next iteration and possibly the overall quarterly plan. As they learn and adjust at the quarterly level, they can provide better fidelity to the annual plan.

Will We Make It?

If you work with leaders, customers, and stakeholders, you will run into the age-old question: How do you know we will make the date?

There is a perception that a traditional schedule provides evidence that we can predict a certain date. We frequently build a highly detailed schedule at the start of the project that spans hundreds of pages. With one of these well-formed schedules we could tell you what we are doing on the last Thursday of October in 2029. Unfortunately this has never been the case in all of our combined years of engineering large cyber-physical systems.

That being said, is all hope lost on increasing predictable outcomes? Absolutely not. Data comes in every day on plan effort versus actual, which can be used to inform our plan. We are collaborating across functions, which identifies unknowns more quickly. And we are leveraging continuous improvement to remove bottlenecks and manage flow.

We can look at the data and make reasonable estimates within ranges based upon the data. Today, if you ask us if we make the date, we would say based on the fact that the teams have been within a 10% variance for plan versus actual, our cumulative flow diagram (CFD) shows optimal lead times, and we have reduced our risk exposure to less than $5,000, we are confident we will hit our date—assuming we do not have a pandemic or other unanticipated events that impede progress.

Communicating Impediments at Saab

Saab Aeronautics in Sweden has shared their approach for communicating impediments.[13] They have a large Scrum implementation with hundreds, possibly thousands, of individuals with a defined escalation path from the individuals on the teams all the way up to the highest levels of leadership. This isn't meant to take away decision authority from the team; however, there are instances where the impediment is beyond what the team can resolve independently. This approach ensures that each person can voice a concern with the confidence it will reach the appropriate level of leadership that same day for action. The goal is to resolve the impediment within one business day, and it becomes leadership's primary task.

Here is their morning schedule:[14]

- 8:30 AM: Executive Action Team
- 8:15 AM: Scrum of Scrum of Scrum of Scrum
- 8:00 AM: Scrum of Scrum of Scrum
- 7:45 AM: Scrum of Scrum
- 7:30 AM: Daily Scrum at the individual contributor/team level

Our CubeSat team is also engaged in daily planning through the use of stand-ups. Individual team members do not have independent stories. The team as a whole collaborates on one story, limiting their work in process, and swarming around the tasks to get the story completed. Each team raises impediments or needs and escalates them to the next level until it reaches the level where it can be addressed.

During the team's stand-up, they are discussing the story "verify that the attitude sensor measures current state and feeds data to attitude navigation." They decide to work on the story collectively using an approach called mobbing, mob work, or sometimes mob programming (Figure 5.12). This is similar to pair programming, a practice made popular by eXtreme programming.

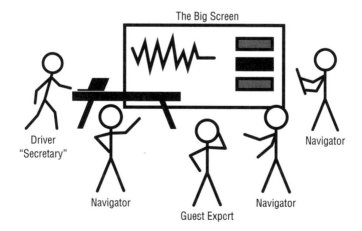

Figure 5.12: Mob Work

During this mob work, all team members are involved as they build and test together. As they are all sitting or standing together, one person starts by taking on the role of the *driver*. The driver sits at the keyboard and begins working the algorithm that will measure the current state of the satellite. The rest of the team are *navigators*. The team observes the work of the driver, provides feedback, and makes suggestions. Through this exercise of daily planning (stand-ups), the team can realize the benefit of improved quality, build a shared knowledge of the complex system, and gain team alignment.

As the teams build the capabilities for their CubeSats (remember, their mission is to bring two hundred CubeSats to launch within the first fifteen months), each Agile team participates in their own daily stand-ups.

Using our CubeSat example, we have a team of eight Agile teams, as shown in Figure 5.13. Each day, the eight Agile teams discuss their plans and share their needs or impediments with each other. First, they discuss if they can resolve the impediment. If not, the impediment rolls up to the next team of teams. This continues until the issue reaches a level that has sufficient span of control to resolve the issue. This approach provides transparency and a shared understanding on how a team's concerns will be addressed as part of a larger, more complex organization.

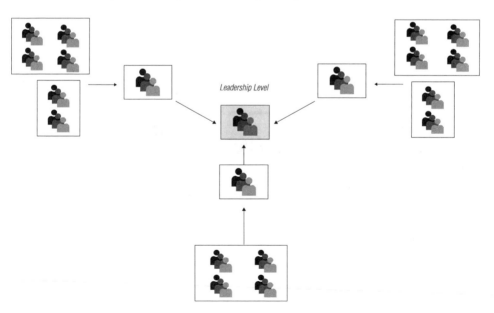

Communication path from teams to leadership. Impediments reach management level every day.

Figure 5.13: CubeSat Team of Agile Teams

CASE STUDY: Saab Aeronautics[15]

There have been numerous conference presentations and publications on Saab Aeronautics Defense's implementation of Scrum for Hardware for their Gripen fighter jet. Their need to build more affordable high-quality fighter jets faster led them to embrace Scrum across software, hardware, and fuselage. They understood that it is not just what you are building that impacts cost, but how you are building it. Improving how teams work can lead to reduced development and production costs.

Over the years, it has been reported that the size of the program has ranged from two thousand to four thousand contributors, with more than a hundred Scrum teams. We know this has been a large-scale implementation, with many individuals building software and hardware components using Scrum. What practices did they employ to help them achieve this goal?

The article "Owning the Sky with Agile: Building a Jet Fighter Faster, Cheaper, Better with Scrum" gives us insight into the ways of working on the Gripen fighter jet. They embraced Agile principles and Scrum along with various development tools to enhance transparency and visibility of progress. One advantage that Scrum offered is the ability to work in short timeboxes, allowing variances to surface quickly, teams to learn quickly, and program risks to be reduced.

They demonstrated multiple horizons of planning with the use of their Strategic Development Plan, with quarterly project plans and sprint-level planning. The use of strategic planning and a project plan provided defined targets for larger releases, such as a specific flight test, yet that plan was decomposed and broken down into quarterly incremental plans, similar to a quarterly road map, with defined milestones. At the execution level, there was a development step that focused on the development and synchronization of activities across the airframe, installation, system development, support systems, etc. and worked through technical risks.

The teams opted for a three-week sprint cycle, and all teams' sprint cycles began and ended on the same day. As teams approached the time horizon, they would break down those large milestones into manageable, actionable, and sensible pieces that would fit within each team's sprint. Saab's teams discovered the need for synchronization (see Chapter 9 for more on cadence and synchronization) at both the sprint level and the quarterly cycle to coordinate and demonstrate progress.

As with all large cyber-physical systems, dependencies are prevalent. At Saab, teams would gather to identify those dependencies and ensure they were visible across the project. Dependencies were reviewed and managed on a regular sprint cadence.

Inherent in an Agile process, continuous improvement is implemented as part of the Scrum framework. Teams were committed to conducting retrospectives to improve the state of the system and processes, and to improve flow and speed of delivery. Continuous improvement is important not only at the team level but also across teams. Therefore, retrospectives were held at multiple levels for process improvements across the system.

Saab demonstrated the ability to build a complex cyber-physical system that is cost-effective, with the ability to deploy a new release of software every six months. According to information published by *Acquisition Talk* in 2021, the Gripen fighter jet costs about $43 million per aircraft.[16] They have been able to successfully demonstrate Scrum applied at scale, across hardware and software, in the development of the Gripen fighter jet.

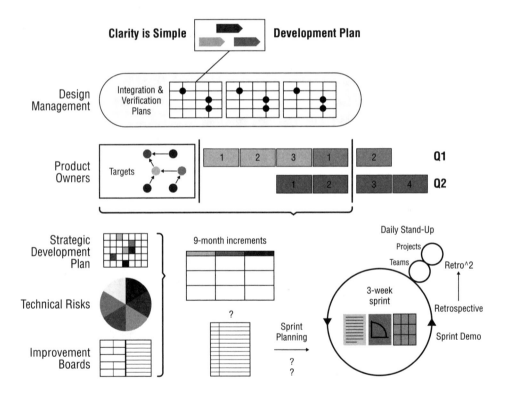

Figure 5.14: Saab Gripen Process

Copyright Joe Justice. Recreated with permission.

Source: Joe Justice et al., "Owning the Sky/SAAB"

GETTING STARTED

- Begin with a vision.
- Develop a Lean UX Canvas.
- Define a high-level timeline based on historical data of products of similar size and scope.
- Define an end-to-end product plan with varying levels of fidelity.
- Decompose time into planning horizons.
- Decompose the system into smaller elements of value.
- Identify and visualize dependency system elements.
- Build a road map for each planning horizon that realizes value.

KEY TAKEAWAYS

- Apply multiple planning horizons.
- Use empirical planning over predictive planning.
- Apply multiple levels of product decomposition.
- Include supply-chain milestones and integration points.
- The further out the plan, the lower the fidelity of the data.
- For large cyber-physical systems, you will typically need six to eight levels of decomposition.

QUESTIONS FOR YOUR TEAM TO ANSWER

- Have you defined and communicated your product's vision?
- How long is your effort estimated to be?
- Have you decomposed the time into timeboxes?
- Have you decomposed your system?
- What dependencies have you identified?
- What are the top customer needs for the next six to twelve months?
- Have you prioritized the system functionality to help create the annual product road map? Have you made the work visible?
- Is everyone who has responsibility for the solution actively engaged in the planning?
- Have you identified your user/customer community, and have you engaged them in the planning?
- Have you engaged your suppliers and aligned them with your cadence?

COACHING TIPS

- Remember, when building cyber-physical systems, there are often many suppliers that need to be considered. Sometimes suppliers are integrated tightly with the product development, and sometimes they are making a delivery aligned with a schedule. Capture those delivery targets as milestones and place them on the product road map for regular review against the critical path. Have well-defined interfaces to allow for more independent design evolution.
- Engage the people executing the work in planning at the quarterly, iteration, and daily planning levels. Those who do the work are the best to plan and estimate the work at the lowest levels. And people are more likely to deliver on their own commitments than commitments made on their behalf. Engage your customers for input into your plans so you have alignment before you get started, and use user feedback to shape the plans. For large cyber-physical systems with many components, sometimes a user might be another system.
- Have an escalation channel from the Agile team level to leadership. Saab Aeronautics has an excellent example of how impediments roll up from the team level to the highest levels of leadership to ensure action is taking place immediately.

CHAPTER 6

IMPLEMENT DATA-DRIVEN DECISIONS

> **PRINCIPLE 3:** Data-driven decisions use current observations and metrics to determine state, manage the flow of work across systems of systems, and continuously improve with real-time data.

Improve your decision-making ability. Be data driven. You are likely thinking this sounds like a great idea. Who wouldn't want to make better, more informed decisions? According to *Forbes*, data-driven decision-making involves using "facts, metrics, and data to make strategic business decisions that align with your company's goals, objectives, and initiatives. It empowers your employees to make informed decisions every day."[1]

But don't we all already use data to make decisions? We have an array of cost and schedule metrics to evaluate progress. Well, therein lies a challenge when building large cyber-physical systems.

For decades, traditional projects have reported progress using well-known predictive planning and management practices—such as fixed scope and fixed schedule with associated analyses models—with progress measured and reported based on phase-gate completion.*

When work is more predictable and repeatable, with few unknowns, creating a mostly static plan can work well. However, that is not the condition of most cyber-physical systems today.

Today's systems are increasingly complex and continuously evolving to meet customer needs or to take advantage of new technologies. Planning happens across multiple horizons and iterations. This iterative approach provides teams with the opportunity to measure progress against

---

* During my (Robin's) time in the defense industry, I had the opportunity to work closely with an officer who had been trained in the special forces. He recounted to me that every time his map of an area did not match the physical terrain of the area, that the terrain was always right. A light bulb went off. He was absolutely right; without objective evidence, information can be out of date or incorrect.

demonstrations of something working versus task completion against a static plan. In data-driven decision-making, the goal is to regularly demonstrate integrated functionality to stakeholders, thus creating frequent feedback loops between development and users.

Demonstrating this type of progress is more challenging in the cyber-physical domain due to physicality; however, embracing digital capabilities along with design and architectural patterns for hardware makes it possible. Making progress visible and seeing results help shape the next set of prioritized work. Visualizing the flow of work across the value stream and supporting metrics of productivity and quality help find the bottlenecks in the system. Progress against objectives and key results helps us understand how we are progressing toward business outcomes, which is more effective than understanding how the individual functional teams are performing. Implementing data-driven decisions focuses on better understanding the progress and state of the product we are building using empirical data and leading indicators, then using that data to make decisions on the priorities for the next iteration of work to improve flow and achieve business outcomes.

In this chapter, you will learn how to use current observations and metrics to determine state, manage the flow of work across systems of systems, and continuously improve with real-time data. This approach helps make the shift from reporting progress based on traditional milestones, such as documentation completed or a list of tasks completed, to demonstrating progress by integrated teams based on objective evidence of something working.

Measure Objective Evidence

With the emphasis on iterative development cycles that embrace the empirical nature of work, we can base reviews and progress not on documentation or tasks completed but on *objective evidence* of something working or something that can be demonstrated against some defined criteria. This approach embraces the concepts of the Agile Manifesto, where we value "working software over documentation." Documentation, such as designs, and models are still needed but are created throughout the development process and are not the focus in how progress is measured.

In the cyber-physical world, we must shift the application of Agile from software-only demonstrations to a systems perspective. As we take a systems view, we begin to ask, "How is the development of the *system* progressing?" Early demonstrations may take place in models, simulated environments, or possibly existing hardware as the new hardware is being defined.

Based on these system-level demonstrations, feedback on improvements or rework, new priorities, or other changes to the direction of the functionality of the system are based on the learning from the last cycle of work and what is observable. These demonstrations enable stakeholders to make better, more informed decisions against what is witnessed. The key is that demonstration of working capability is the *true* measure of progress.

As an example, let's take a look at our CubeSat development team. They are currently focused on developing the algorithm to *determine* the attitude so we can *adjust* the attitude when needed. As they build out the functionality, they need to build the algorithm and verify through demonstration that it is functioning as intended (see Figure 6.1).

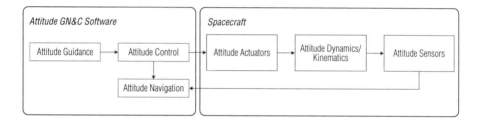

Figure 6.1: Attitude Guidance, Navigation, and Control
Source: Steve Ulrich, professor at Carleton University in Ottawa, Canada, and the founding director of the Spacecraft Robotics and Control Laboratory.

But suppose we identified that our current communications system was inadequate for the CubeSat: the current range is five hundred kilometers, and we need ten thousand kilometers. The team may decide to plan, design, and implement a feature for improved communications. The work executed by the team may include updated requirements, designs, hardware, and software. However, the objective evidence we want to measure progress against is the demonstration of the updated communication range systems. To validate the progress, the team transmits a signal to and from the CubeSat and the ground system, which would measure the difference in signal strength, data error rate, and latency change through an algorithm such as a link budget calculation. In this example, our measure of progress is the change in communications range, not how many artifacts were completed.

As the team plans their work, they discuss the functionality, how it will be tested and demonstrated, and what environments or digital capabilities will be used. For the demonstration, they use a software simulation for the attitude controller and demonstrate the new functionality in a simulated environment and with the evolution of a digital twin (see Table 6.1).

Table 6.1: Example Backlog with Objective Evidence for Demonstration

Backlog Item	Time Horizon	Description of What is Needed	Objective Evidence and How It's Demonstrated
Epic	>1 Quarter	Implement attitude controller	Incorporate into hardware-in-the-loop closed-loop simulation (e.g., partial physical twin).
Feature	Quarter	Determine attitude so that we can adjust the attitude.	Incorporate into the target software environment; demonstrate on a processor-in-the-loop environment/simulation—hybrid—cyber-physical twin, emulated processor/other subsystems. Demonstrate the adjustment made in the attitude controller.
Story	Iteration	Verify that attitude sensor measures current state and feeds data to attitude navigation.	Incorporate into software-in-the-loop simulation for the attitude controller. Demonstrate with the digital twin/model (the digital twin is still being built out).

As the team demonstrates new functionality as part of their defined iterative cadence, if it all works according to the acceptance criteria and test plan, then all is good: no changes or adjustments are needed. Alternatively, the demonstration might show that the calculation is slightly off or not displaying results as expected. Regardless of the outcome, these demonstrations provide real-time, objective evidence of how the new functionality is working. Through this analysis, the team and their stakeholders can determine the current state of the product and use that information to determine the next steps of development. Next steps could include new functionality, reprioritization of functionality, or improvements with the existing functionality.

As we apply multiple horizons of planning, each level of the backlog is defined to understand how the functionality is demonstrated. Table 6.1 provides an example of what this might look like at each level of the product hierarchy. The demonstration provides the observable, objective evidence of product progress and the opportunity for feedback from stakeholders as the integrated system is built.

The early demonstrations occur using simulations and models and hardware-in-the-loop environments, where the software is tested in an environment closer to where the software will be, improving reliability. For example, the team might demonstrate in their digital model the impact of a positioning algorithm on the spacecraft and receive feedback on their next steps in development. These fast feedback cycles reduce variation, which, in turn, reduces costs and keeps us closer to delivering on schedule.

Feedback cycles provide regular validation of the system functionality to ensure teams are building the right solution while providing the opportunity to engage the customer and the business team in the process. The teams also learn through the demonstrations how the system is performing and will identify where refinements are needed.

Based on what is observed, the team learns whether small course corrections need to be taken. Don Reinertsen, author of *The Principles of Product Development Flow: Second Generation Lean Product Development*, informs us that short iterations with fast feedback based on objective evidence of something working can lead to improved economic value. As Reinertsen points out, the advantage of short iterations of a couple weeks is that only small variations and course corrections occur, whereas larger cycles lead to greater variances and major schedule delays.[2]

This is quite different from the traditional waterfall predictive approach, where integrated demonstrations are few and far between, creating the potential for increasing variance between what functionality is desired and what is being developed, as shown in Figure 6.2 (page 86). Documentation and traditional milestones are accepted as value delivered, yet there is no demonstration of working functionality, and the system is not going through iterative validation. In the case of cyber-physical solutions, even small deviations can result in huge monetary losses or time delays.

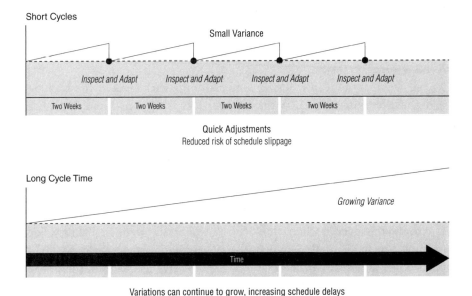

Figure 6.2: Iterative Development Provides Fast Feedback and Decreases Deviations

Digital Tools and Engineering Environments

Having a backlog of capabilities with defined demonstration criteria is only part of the journey to data-driven decision-making. With cyber-physical systems, the act of small iterative development and demonstrations is made possible only when you have instrumented your system to provide data and leverage digital tools and engineering environments. These digital capabilities result in technical agility and reduce the cost of change for cyber-physical solutions.

Often, the physical hardware is not available, especially early in the product development life cycle. However, thanks to the latest technologies and surge of digital transformation efforts, we now have opportunities to develop, test, and demonstrate iteratively using a variety of tools. As the new hardware is being defined and developed, it can be tested using 3D printing options, prototypes, digital and simulated models, digital shadows, digital twins, and the use of emulators.

At Saab Aeronautics Defense in Sweden, hardware development teams make use of digital environments, and each developer has access to the models and simulated environments so they can test and see the results as they develop new features. They can make use of the simulated environments any time with flight simulation running their current CAD and code on their individual machines. Their simulations enable them to visualize the impacts of their updates, such as the addition of a new fuel pump or a new bolt. They can also use the simulated environment to fly the plane around as they study updates to the system, or they may want to see what happens if there is an interference with the plane and use this data as input for the next planning cycle.[3]

While designing and building the next-generation Gripen fighter jet, Saab tested new radar features using the earlier version of the Gripen to get some initial feedback on how the new design and features were working. Their iterative approach and use of models created the opportunity for regular synchronization of product enhancements and demonstration of new functionality.[4]

Visualization Tools

There are many visualization tools and techniques available that provide data to visualize the progress of the work performed, the current state of the product across different steps in the value stream, the impact of new functionality to the system, and test results. Using the data provided by these tools, management has a more accurate understanding of how the product is progressing across the value stream, such as development, supply-chain status, and the factory floor.

The importance and impact of visualization have been known in training, education programs, and engineering for decades. Edward Tufte, a forerunner of data visualization, was an early advocate for the use of visualization and how it can impact the decision-making of management. Visual displays and data help move us from subjective and sometimes emotional decisions to objective decisions based on quantifiable data points and trend analysis.

Digital Models

Models, simulations, and digital twins provide another perspective for making data-driven decisions. We capture the data over time, providing us with a digital thread. According to the *AIAA* article "Engineering Design with Digital Thread," a digital thread provides the traceability of "data-driven architecture that links together information generated from across the product life cycle."[5] The models of the system are updated and integrated, providing real-time data on the status and configurations of the system. The integration of these models provides a digital copy of the system configuration, ensuring quality standards are met. As the system goes through development and manufacturing, real-time data can be used by management, engineering, schedulers, and those on the factory floor to have visibility into the state of the product, the impacts on the system as changes are made, and reallocation of resources when bottlenecks in materials arise, and they can use this to improve the product and the business processes surrounding it. In their paper "Visualizing Change in Agile Safety-Critical Systems," Cleland-Huang and colleagues state, "By leveraging traceability, we can visualize and analyze changes as they occur, mitigate potential hazards, and support greater agility."[6]

A digital twin is the digital replication of the physical system. It receives information from a variety of sources updating its current configuration and performance status. This real-time data provides the ability to understand changes in the deployed system, such as performance degradation or the need for part replacement. In addition, the digital twin is used as a simulation environment to test changes in the digital environment first before releasing to the physical product. This helps to reduce risk and avoid potentially costly mistakes. As Somers and colleagues emphasized in their literature review of digital twin–based testing for cyber-physical systems, the importance of digital twins is to provide "data-driven and simulation-based models coupled to physical systems to provide visualization, predict future states, and communication."[7]

Tesla is a prime example of how digital twins can be used to collect data to improve and sustain their products. During a discussion at a workshop we had with Joe Justice in the fall of 2022, he explained that the use of digital twins is commonplace at Tesla, as a digital twin is created for every car made. Data is collected from the physical vehicle and sent to its digital replication. The data analysis helps Tesla understand the current system state of the vehicle and where or what repairs or maintenance might be needed soon. When Tesla uses this data with predictive analysis, updates to the vehicle can be made before it breaks down, providing a better user experience for the owner. This also benefits Tesla and their desire to have satisfied customers and a growing business.[8]

—

Through digital transformation and emerging technologies, iterative development with regular demonstrations and real-time data analysis have become a reality in the development and production of cyber-physical systems. Those companies that have taken this leap have recognized that the use of these modern techniques enables faster learning cycles, resulting in demonstrated learning and improved decision-making. This provides an enhanced decision-making framework with real-time data, enabling organizations to continuously improve their products and reduce time to market.

Measures for Value Management Office

In the book *From PMO to VMO: Managing for Value Delivery*, the authors state, "Metrics bring strong focus to process, to outcomes, and ultimately to behaviors. Chosen correctly, metrics will foster and accelerate agility."[9] Demonstrations are effective in understanding the current state of the product. The iteration cadence (multiple horizons of planning) provides the time to inspect and analyze results and provide feedback to the team for the work recently completed. It is time to discuss value delivery through demonstration.

However, empirical evidence is only one form of input into the data analysis and decision-making processes. There are other questions that need to be answered, and additional data points are necessary to understand how we are progressing toward the business outcomes.

Business Outcomes

Business outcomes are typically lagging indicators, such as improving time to market, improving operational efficiency, and increasing customer satisfaction. They are often measured over a significant period (a year or more); therefore, it is necessary for a business to embrace nearer-term measures that provide inspection points toward the desired business outcomes.

Objectives and key results (OKRs) used annually or quarterly connect business outcomes to the execution of work done by the teams. OKRs create alignment around shared objectives with clear, demonstrable progress on a regular cadence. Based on the work of John Doerr, the objective sets the direction the organization wants to go, with key results being the measurable and observable milestones.[10] As key results are demonstrated as the product is developed, the organization gains understanding on whether they are achieving the desired outcomes.

Desired outcomes vary greatly across organizations based on their strategic needs while driving toward some defined delivery of value. For example, the business outcome for an organization in the space industry might be a successful satellite launch by a target date as part of a growing market strategy. For an organization in the automotive industry, their business need might be to increase customer satisfaction through improved predictive analysis and responsiveness. Therefore, they might focus on an upgrade to their maintenance and diagnostic systems targeted for a release in the next twelve months.

Because of the long lead times in achieving such business outcomes, each quarter has a set of defined objectives and key results to assess how they are progressing toward reaching the target. OKRs are aligned to team-level work, and the teams' demonstrations give insight into the progress of the OKRs, providing a clear understanding of how the team, and the overall integrated system, is progressing and where new opportunities might be emerging. The OKRs are a set of leading indicators that provide metrics at the value-stream and integrated-system level.

OKRs are one form of measure used to gain visibility into the state of the product toward achieving business outcomes. There are other measures necessary in managing and improving the state of the product. The data from numerous sources can help identify bottlenecks so improvements are made at the right place within the value stream in order to ensure safety, quality, and maintainability and improve speed of delivery and other metrics to improve decision-making based on real-time data and visual displays.

Measuring the Flow of Value

With the surge and progress of digital transformation efforts, what may seem like a game changer now is quickly becoming table stakes. Industrial DevOps recommends a variety of metrics and opportunities to enable transparency of the state of the product across the entire value stream, from design, development, and manufacturing to operations and customer services.

In *Project to Product*, Kersten advocates for organizations to "create a feedback loop between flow metrics and the business outcomes that they generate, which then creates the continual learning and experimentation loops that enable high-performance [IT] organizations to thrive."[11]

This feedback loop provides ongoing validation of system capabilities. In addition, organizations use a variety of metrics to understand the progress of the development and manufacturing of cyber-physical systems. The measures we outline below are based on well-known works in the industry, such as *Project to Product* by Kersten, *Actionable Agile Metrics for Predictability* by Daniel Vacanti, DORA's *State of DevOps* reports, and Lean-based measures. These are leading measures that help demonstrate the path toward value delivery to the customer and achieving business outcomes, as shown in Figure 6.3.

Leading Indicators

Measures when driving toward a desired business outcome

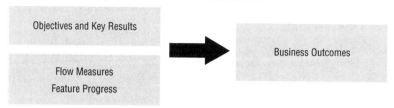

Figure 6.3: Leading Indicators of Business Outcomes

These measures have been proven in the software industry and apply in the cyber-physical domain as well. For metrics to be useful and effective across teams of teams requires integrated digital environments and standard metrics collection. Based on the ISO Update website, having a set of defined standards within an organization enables them to "effectively share their goals, processes, procedures, and vocabulary needed to meet the expectations of their stakeholders."[12]

Table 6.2: Measures of the Flow of Value

Measure	Description
Flow time	Measures time to market; namely the time elapsed from "work start" to "work complete" on a given flow item, including both active and wait times."[13]
Flow efficiency	Is the ratio of active time out of the total flow time.
Flow velocity	Is the number of items being completed over a defined unit of time. In our context, flow velocity is measured using story points and is the range of story points a team delivers over several iterations.
Flow load	Measures the number of flow items currently in progress (active or waiting) within a particular value stream.
Flow distribution	Measures the distribution of the four flow items—features, defects, risks, and debts—in a value stream's delivery.
Work in progress (WIP)	WIP can be thought of as the functionality, architecture, or inventory work that is in progress but not yet completed.

Flow measures (in Table 6.2) are seen in most Lean and Agile communities and emphasize the need for digital capabilities to make progress visible. Through automation, these measures can improve the flow and delivery of value to the customer.

Using the definition by Kersten, flow time is defined as the "time elapsed from when a flow time enters the value stream (flow state becomes active) to when it is released to the customer or when it is launched (flow state = done)."[14] Flow time is *like* the Lean community's concepts of lead time and cycle time. As we explore what these measures mean for cyber-physical systems, let's start with some well-defined definitions from the industry (see Table 6.3), and then we will build upon these with an example (see Figure 6.4).

Table 6.3: Lead Time vs. Cycle Time

Measure	Description
Lead time (production lead time)	"The time required for a product to move all the way through a process or a value stream from start to finish. At the plant level, this often is termed door-to-door time. The concept also can be applied to the time required for a design to progress from start to finish in product development or for a product to proceed from raw materials all the way to the customer,"[15] as defined by Lean.Org/Lean Enterprise Institute. Lead time is the period between the start of a process and its conclusion. That is, it's the amount of time it takes to make a product or service so it's usable for the customer.[16]
Cycle time	"Time required to produce a part or complete a process, as timed by actual measurement,"[17] as defined by Lean.Org/Lean Enterprise Institute.

Lead time has the longest measure of time, going from customer need identified through to delivery. Flow time is a subset of lead time and starts from the time the work item (feature or story) becomes active. Cycle time is a smaller unit of time that looks at a specific process within flow time. It could be looking at the cycle time between different steps or a specific process within the development value stream.

With large cyber-physical systems, there are multiple levels and components that come together to deliver the end system, and there are value streams, nested value streams, and development value streams. Going back to our CubeSat mission, we have a satellite value stream and a development value stream. Based on this structure, there is the lead time for the entire satellite value stream and for the attitude control system, as shows in Figure 6.4. We could measure the flow time for the story "verify that attitude sensor measures current state and feeds data to attitude navigation."

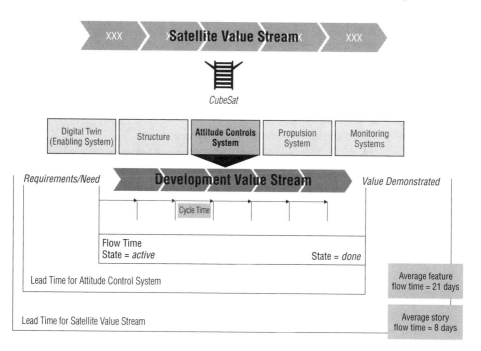

Figure 6.4: Flow Time Illustrated

To measure the story's flow time, you would start counting once the story became "active"—that is, once work has been started—and would stop counting once the story was accepted as "done" and demonstrated in a system environment. Flow time can also be used to measure the time it takes for a *feature* to go through the development value stream. Lead time would measure how long it took for the feature (such as a satellite enhancement of a capability) to be released to the spacecraft.

Whatever the scenario, it is important to understand how long it takes from the start state to delivery. Management might use this data to understand if the deployment of the system is meeting the desired return on investment and delivering value soon enough to meet customer needs. If the goal is to deliver in the shortest sustainable lead time, then this measure is important to understand if improvements implemented (tools or processes) are improving the flow time. Measuring flow time helps understand how long it is taking for each work item—for example, a feature—to get through the process. Improving processes can improve the overall flow time.

While flow time provides data to understand how long it is taking to deliver some functionality or end-to-end capability, flow velocity is the number of items being completed over a defined unit of time.[20] In cyberphysical systems, flow velocity using story points measures the range of story points a team delivers over several iterations. A simple example is from one of the CubeSat teams. Let's say that over six iterations, they can, in general, complete forty-eight to sixty stories. They find their average velocity is fifty-two points. They now have a historical number they can use to forecast the completion of future work (see Figure 6.5).

When a team understands their velocity, they have improved team predictability and higher confidence in meeting their objectives. Measuring flow velocity provides historical data that can be used to understand, and hence predict, how long it will take to deliver future work items. This aids teams with the planning and alignment of work when many teams are engaged in delivering integrated functionality. While the example shows one team, this approach can be used across teams whose collective work must be integrated to understand how the system is progressing with feature completion toward a defined release, like a flight test for an airplane or a launch date for a spacecraft.

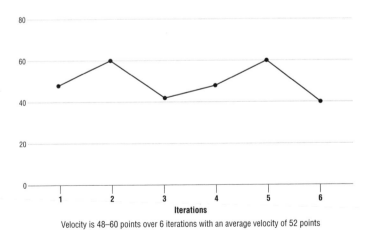

Velocity is 48–60 points over 6 iterations with an average velocity of 52 points

Total story points/Number of Iterations = Average velocity

Figure 6.5: Flow Velocity for a Team

While the desire is always on reducing dependencies between teams, when building large cyber-physical systems, there are integration points with multiple software components, embedded software, and hardware as well as integration with multiple suppliers. Improved predictability can help large-system teams better understand their ability to achieve a defined release date or integration point.

DORA's *State of DevOps* research has identified four core metrics we should also consider.[21] While this research applies to software systems, we can apply the metrics to cyber-physical systems. In cases where the product has not been deployed to its operational context, we can substitute high-fidelity simulated test environments:

- **Deployment frequency:** how often a software team pushes changes to production
- **Change lead time:** the time it takes to get committed code to run in production
- **Change failure rate:** the share of incidents, rollbacks, and failures out of all deployments
- **Time to restore service:** the time it takes to restore service in production after an incident

Cyber-physical systems present unique challenges with testing, as they span across software, hardware, and electronics. Simulated environments are critical for testing new functionality and gaining early visibility into the impacts of the changes on the system, along with testing functionality in hardware. Research performed by Somers and colleagues in "Digital-Twin-Based Testing for Cyber-Physical Systems" suggests that "hardware-in-the-loop testing be used to test specific physical components of a system while simulating the remainder of the system."[22]

Visualize Work in Progress

It is important to visualize the amount of work in progress (WIP). WIP can be thought of as the functionality, architecture, or inventory work that is in progress but not yet completed. Just like on the freeway, when the traffic (WIP) is too high, everything slows down, and it takes longer to get from point A to point B. If your teams want to improve flow time, you need to understand and limit their total WIP.

Visual tools, such as Cumulative Flow Diagram and kanban boards, make it easier to monitor WIP and see when there are bottlenecks. When there is a bottleneck (flow has ceased), it becomes visible, and action can be taken to resolve the issue. Remember our story about Saab Aeronautics from the previous chapter? They have a built-in escalation path from the team to senior management, where everyday teams can elevate their impediments so action can be taken right away by the appropriate level of ownership. The goal is to resolve the impediment within twenty-four hours so flow in the manufacturing line can continue.[18]

CFDs are a visualization model that can portray a variety of information that helps teams understand the state of specific work items over a given time frame and if the items in each state are growing or reducing. For example, a CFD could be looking at the state of team-level features across several months to determine how long features are staying at different stages of development through to delivery.

According to *Actionable Agile Metrics for Predictability*, to improve and understand system predictability means understanding the actual process performance.[19] The CFD indicates where features may be impacted by a bottleneck in each development state, such as systems testing. When that occurs, teams can investigate what is occurring in the process and what needs to be improved.

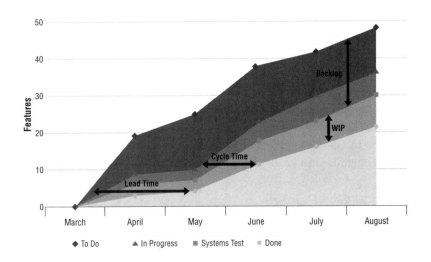

Figure 6.6: Cumulative Flow Diagram

For example, the data could be showing a growing bottleneck (as illustrated in Figure 6.6), which is increasing WIP at the systems test stage and increasing the flow time. Once the situation is identified, steps to improve the situation can be taken. In this case, it may mean improving the testing environment, or maybe it means reducing the number of items in development and having the team focus more on the testing to get features completed. The CFD provides the data so management can take the next steps in analyzing the situation toward a resolution.

Feature Progress

Feature progress helps the customer and business owners understand what percentage of feature has been completed based on story acceptance criteria completion. By understanding feature progress, business owners can anticipate when the feature will be ready for demonstration and release to the customer.

The CubeSat team is building the following feature: develop attitude error estimator and integrate into the processor-in-the-loop (PIL) environment. The feature is composed of user stories, which implement the features acceptance criteria. With each demonstration, we evaluate the percent completed of the feature acceptance criteria.

If we have completed five out of ten items in the acceptance criteria, assuming all are weighted equally, we would be 50% complete. If teams want even greater fidelity on percent complete, they could weigh the size of each story used to implement the criteria.

How Measures Work Together to Make Better, More Informed Decisions

In the section above, we outline a set of measures that work together to inform data-driven decision-making. But what does this look like in practice, and how are the resulting metrics used as part of the process?

You may recall the CubeSat team's mission is to improve weather forecasting accuracy through better data collection. Imagine they are in the process of building the CubeSats, and they received a launch date for the first round of CubeSats, which is fifteen months out. They have been working for a few months and are having a demonstration and review with their stakeholders to get feedback on their progress. During each iteration, the CubeSat teams demonstrate new functionality and use one or more objective measures to communicate progress toward a milestone or concerns.

Business outcome: To launch CubeSats on June 3 from Wallops Island.

Their primary objective and key results include: Demonstrate attitude control system functionality of the CubeSat as measured by (1) demonstrating the current state based upon sensor measurements and (2) the desired attitude state in a simulated environment and hybrid cyber-physical twin.

It is the end of the iteration, and now the team will demonstrate that the attitude sensor measures the current state and feeds data to attitude navigation through a software-in-the-loop simulation for the attitude controller in a simulated environment with the ability to measure the current angular velocity of the spacecraft.

The steps the CubeSat team will take in the iteration demonstration include the following:

- The team demonstrates, in a simulated environment, and verifies that the attitude sensor measures the current state and how it feeds data into the attitude navigation. At the story level, this demonstration is performed via a software-in-the-loop simulation. The stakeholders (internal stakeholders or customers as desired) participating in the demonstration provide feedback on the demonstration, responding to questions like these: Is the functionality working as intended? Is the data displayed accurately? Progress toward the objective and key results are also discussed for feedback from stakeholders and whether the state of the current functionality is on track to achieve the OKRs for the quarter.
- The team then shares some data regarding the state of the processes and predictability measures.
- The team also discusses their flow velocity. Their flow velocity over the past six iterations is an average of fifty-two points. They have three iterations left in this quarter and currently show 150 points of remaining work in their backlog. They discuss it as risky, but if they perform at fifty points per iteration and do not experience any major bottlenecks, they should be good. There are no major holidays during this time. The architect sitting in

the demonstration is thinking about some security features they should consider building into this quarter. Obviously, based on the data, the team is not able to take on more work than their capacity allows, so they begin a discussion. Should they reprioritize work for the remainder of the quarter? Should the security work be the priority in the next quarter? The data is valuable in supporting their decision-making process.

- Next, the CubeSat team shares the CFD. The current CFD shows that they have a growing backlog of work not started. This is a result of the new security requirements they recently received and is currently being defined in the backlog. Recently, some defects have been identified. The growing backlog is a concern since they now have a targeted launch date. They use data to help them figure out how much risk they are incurring. Based on their current velocity and the growing backlog, will the team be able to handle the work? Do they need to engage another team? Is everything in the backlog necessary for launch? While the data from the CFD is highlighting a growing backlog, the teams compare the estimated work in the backlog against their point velocity. Because of the CubeSat's digital engineering environment and integrated tools, they are able to use real-time data to support their decisions and the next actions they decide to take.

Feature progress: As the CubeSat teams build their constellation, they will also demonstrate and validate that the system is working as intended. They continuously inspect and improve the system and processes to build features faster. They inspect their progress using their integrated working environments. Having the data is important. Knowing how to use the data is vital in implementing data-driven decisions.

CASE STUDY: NASCAR Advances Race Car Development through Simulation[23]

NASCAR was founded in the late 1940s by Bill France Sr., who was a mechanic with a passion for stock car racing. To this day, NASCAR races are among the most watched programs on television. In fact, in 2023 the Daytona 500, one of NASCAR's premier races, sold out for the eighth year in a row. NASCAR is not only entertaining but is also a case of where product development is continuously improved through making data-driven decisions.

NASCAR announced they needed a new generation of race cars in 2018. Designing a new race car is both complex and challenging due to regulatory requirements, frequent rule changes, aerodynamic testing, and performance optimization. Typically, a new race car will go through thousands of hours in expensive wind tunnel testing as it's being developed. In addition to the design, NASCAR teams have to continuously improve. NASCAR races almost weekly during the season, and their teams typically have less than one week to prepare in between races. Unfortunately, they are on a new track every week and often will have fewer than two practice runs. The recent advance in simulation provided by D2H and Ansys has improved the design of the cars and streamlined aerodynamics in cars to nearly eliminate wind tunnel testing.

NASCAR uses instruments on their vehicles to provide full telemetry of RPMs and other performance metrics. They use sensors to track surfaces, weather conditions, and competition rules. They use the data to run exhaustive computer simulations that enable teams to produce three times as many designs without extra development time and resolve issues in minutes to hours. The data-driven approach to decision-making can equate to millions of dollars in purse and sponsorship money weekly.

GETTING STARTED

- Define your yearly OKRs for alignment between strategy and execution. It will help the teams understand the direction they are headed, focus priorities, and build connectedness.
- Review the flow of value through the system. To do this analysis, you need to look at the tools. How quickly are features getting completed? Where are the bottlenecks in the flow? Once you find the bottleneck in the system, what will you do?
- Define a small set of metrics. As you review each of the measures, which ones will you start with to help drive the outcomes you are after?
- Objectives are fulfilled by features. Each feature needs defined objective evidence. Capture with each feature how it will be demonstrated.
- Invest in digital integrated tools to improve visibility and progress with real-time data. Many organizations are going through some level of digital transformation, and one area to focus on is the integrated tool environment. Invest there.

KEY TAKEAWAYS

- Measure progress based on objective evidence of work demonstrated, not on tasks completed. The goal is to regularly demonstrate integrated functionality to stakeholders. This is more challenging in the cyber-physical domain with software and hardware, but embracing digital capabilities and design and architectural patterns for hardware makes it possible.
- Making progress visible and seeing results help shape the next set of prioritized work.
- Visualize the flow of work across the value stream to help find the bottlenecks.
- Progress toward meeting objectives and key results is understood as features are demonstrated. For cyber-physical systems, this is especially challenging, as we measure flow across the value stream, which includes software, hardware, and often suppliers.
- Stakeholder participation and feedback at the demonstrations are critical in validating the solution is meeting the anticipated results.
- Digital tools and engineering environments are enablers for iterative development and demonstrations with cyber-physical solutions.
- There are a variety of metrics that can be used. Know what metrics you are using and why you are using them. Ask yourself if these metrics are helping you understand your progress toward business outcomes.

QUESTIONS FOR YOUR TEAM TO ANSWER

- Has your organization captured the value stream for the product? Before you start iterating on designs and development, be sure the teams are organized around the value stream (Principle 1).
- Once you understand your value stream, document the different development environments. When teams do their demonstrations, which environment will the demonstration take place in? What tools are needed?
- Has your team identified which metrics to track and how they will use the metrics to make data-driven decisions?
- Which tools do you need to obtain quantitative data?

COACHING TIPS

- Basing decisions on objective evidence requires regular demonstrations of integrated functionality.
- Complete a stakeholder analysis to ensure you have the right participants at the reviews to provide feedback and assist with and improve decision-making.
- Reduce risk and schedule delays through regular observation and solution demonstrations.
- Be intentional in your digital transformation journey by prioritizing your digital capabilities against the greatest return for your needs and goals.

CHAPTER 7

ARCHITECT FOR CHANGE AND SPEED

> **PRINCIPLE 4:** Architecting for change and speed provides information on multiple architecture considerations that can reduce dependencies and improve the speed of change.

Good architecture leads to better systems. Having a clear blueprint of what we are going to build before we begin is clearly a good system. It allows teams to effectively communicate the needs and dependencies of the system to those who will be building it. The more complex the system, the greater the need for an easy-to-follow blueprint, especially when that system is really a system of systems comprising software and firmware/hardware. There is a common misunderstanding that we do not need architecture for Agile or DevOps teams, but nothing is further from the truth. Architectural artifacts are used for many purposes when building cyber-physical systems, which include providing a common framework to enable design, a mechanism to communicate the design to stakeholders, a benchmark to verify and validate the system against, and a baseline to maintain and evolve the system in the future. Our goal is to build a blueprint that supports the needs previously discussed while ensuring we can adapt to change rapidly.

But how do we architect complex cyber-physical systems and still develop better systems faster? How do we get to market at the speed demanded of organizations today, without sacrificing quality or safety? How do we achieve the goals of continuously developing, integrating, deploying, and releasing value at speed in complex cyber-physical systems? The key is to architect for scale, modularity, and serviceability.

In this chapter, you will learn what Industrial DevOps architectural considerations you should consider in designing and building cyber-physical systems, what current technology trends are impacting architecture, and how you can leverage those trends as architectural accelerators (i.e., architect for speed). (We are not trying to teach general rules of good architecture but will focus on the specific Industrial DevOps solutions that help cyber-physical systems architect for speed. For those of you who have less of a background in systems engineering and architecture, you can first go to Appendix B to learn a little bit more.)

Architecting Cyber-Physical Systems

No matter what you are building, it is important to have an intentional road map of the required elements when architecting your system, because there are so many trade-offs that need to be considered, including change, usability, availability, observability, agility, manufacturability, reusability, security, and scalability. Each of these main considerations is further complicated by sub-considerations. The considerations for architecting cyber-physical systems are outlined in a road map in Figure 7.1. By focusing on each of these considerations and making good architecture choices up front, we can avoid costly delays and build better systems fast.

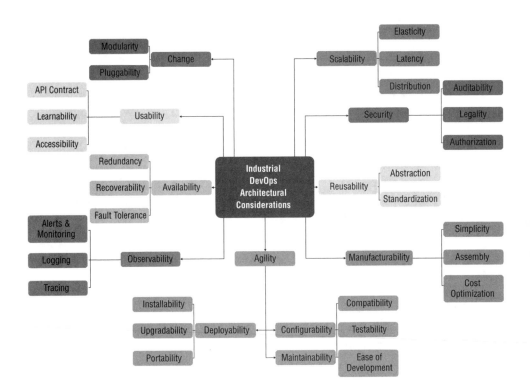

Figure 7.1 Architectural Considerations for Industrial DevOps

In many cases, engineers start the design process of a new system by copying an existing system's architecture and then applying it to the next system. This is done to save time in architecting a new system from scratch and is a practice used by both software and systems engineers. The problem comes when we don't stop to ask if the mission or purpose of a new system is the same as the one we are copying. We must take the time to stop and consider the unique needs of this new system. Context matters.

Teams should first start any new systems architecture by creating a quick table listing the architectural considerations for the system, then rate the impact they have on the new system using something like a Likert score. This is a very quick exercise that highlights areas in the architectural pattern you are using that may need to be refactored. Let's walk through several of the major architectural considerations for cyber-physical systems. By planning for each of these up front, we can architect a system for speed.

Architect for Change

Given the rate of change in technologies we are experiencing, it is critical that teams building cyber-physical systems begin by architecting for change and extensibility. Many systems morph throughout the development process, sometimes eventually providing capabilities that they were not intended to provide. When this happens to a system that was not built for change, the result can be a Frankenstein's monster of work-arounds.

Creating an architecture that is built for change involves designing a system that can be easily modified or adapted to meet the evolving business needs, new technology trends, and changing customer requirements. Key attributes to architect for change include *modularity*, which is dividing the system into small independent modules, and *pluggability*, which allows you to easily replace these modules. When architecting for cyber-physical systems, there are several unique challenges, such as heterogeneous interfaces, where we mix hardware and software components from a variety of suppliers and need to ensure that they communicate.

A good example of a cyber-physical system that was defined with change in mind is Lockheed Martin's SmartSat technology. The smart satellite can be reprogrammed in orbit for entirely different missions. During my (Robin's) time at Lockheed Martin, executives described the technology as a smartphone in space, which I thought was an excellent metaphor. Smartphones are very powerful computers that can operate as many different devices, like a camera, navigation system, calculator, or a phone. By downloading a new app, you can add entirely new capabilities to your phone. The smart satellite can do the same. The satellite is a module with pluggable capabilities that connect through standardized interfaces.[1]

The smart satellite technology is a software-defined, hardware-reliant system where the hardware is configured and controlled by elements of the software. The software-defined, hardware-reliant system is a growing trend across all industries and is often known as *software-defined everything* (SDE). SDE uses software to abstract and control the hardware for computing, storage, networking, and physical devices, which provides a solid foundation for system agility.

To build for change, begin by understanding the drivers of change, such as potential business needs or technology trends on the horizon. Modular design and applied design patterns can help teams build in flexibility. It's also important to identify areas that may impact scalability, such as data growth or user traffic. Teams should run experiments with hypothetical changes and validate the areas that have the least flexibility.

Architect for Usability

Usability is critical to building better systems faster but is often not given enough emphasis. In the cyber-physical world, usability is much more than the aesthetics of a system; it can also determine safety for the humans in those environments. Take the cockpit design of an aircraft. Architecting for usability is a critical consideration, as a clear and intuitive user interface can help reduce the risk of pilot error and improve flight safety.

When architecting for usability, it is important to account for learnability, efficiency, memorability, failures, and satisfaction. During product design and development, it is important to focus on simple intuitive interfaces using metaphors and consistent design patterns and visual elements. Vehicle rear view cameras are an example of an intuitive interface in a cyber-physical system. The camera provides not only a visualization but also guardrail lines that show where the vehicle is in relation to other objects.

Architect for Availability

Availability refers to the ability of a system or service to be accessible and operational. Cyber-physical systems typically have high availability requirements. For example, cars need to be accessible and operational every time they are in use, or the consequences could be fatal. Key attributes to consider in this area are redundancy, recoverability, latency, and fault tolerance. While these areas have always been considered in architecting cyber-physical systems, new technologies have changed or enhanced how we can accomplish this.

Trends in cloud computing have enabled new ways to achieve redundancy and recoverability, which are achieved by automatic replication of data across multiple datacenters and load balancing. Edge computing, which processes data closer to the edge, has provided a mechanism to reduce latency. Chaos engineering (the discipline of experimenting on a system to build confidence in the system's capability to withstand turbulent conditions in production) has improved our approach to fault tolerance by injecting faults into the system randomly, forcing an improvement in resilience. Today, NASA uses chaos engineering in simulators that regularly inject failures into the system, greatly reducing risk in actual launches.

Architect for Observability

Observability has always been a concern, but the term *observability* in association with software-intensive systems is relatively new and really came into focus with the rise of DevOps and the goal of continuous feedback. In addition, the adoption of Industry 5.0 principles has resulted in the need for increased visibility into system performance and health. Instrument systems need to provide full telemetry regarding behavior and performance of the system.

Today, technological advances help us build observability in new ways. For example, when you order dinner on Uber Eats or DoorDash, a screen pops up that shows your order being made, the driver picking up the order, and the delivery of the order to your house. The user has full observability into the system. That same level of observability into a cyber-physical system, such as a car, is very helpful in identifying the current state of the system, understanding where updates need to be made, identifying where problematic behavior is coming from, and even alerting the user of intrusions. Key attributes to consider for observability are alerting, monitoring, logging, and tracing.

New technologies are not only driving improvements in greenfield systems. In September of 2020, the DoD installed Kubernetes on the U-2 Dragon Lady test flight.[2] The flight computers on the U-2 were able to use Kubernetes to run advanced machine-learning algorithms without any impact on the aircraft's flight or mission systems. This allows us to capitalize on the aircraft's high-altitude line of sight and makes it even more survivable in a contested environment.

Architect for Scalability

Systems need to be designed to easily adapt to the changing needs of users and stakeholders. Scale is not just about growth; it's about the system being able to adapt. For example, a manufacturing system that uses robots to assemble products needs to be able to dynamically allocate robot resources to different assembly lines based on demand. Key attributes to consider for scalability include elasticity, latency, and distribution.

Elasticity is the system's ability to dynamically adapt resources based on changing needs. Cloud technology has increased our ability to improve elasticity through distribution, where we can spread workloads over multiple resources. Advancements in areas such as in-memory computing, high-speed memory, and faster processors reduce latency concerns.

Formula One has taken architecture for scalability to the next level by leveraging Amazon Web Services (AWS) to scale their computational fluid dynamics technology environment, which allowed them to use empirical data to design their 2021 car. The environment enabled them to run 1,150 compute cores and analyze more than 550 million data points that model the impact of one car's aerodynamic wake on another to pick the optimal design. The AWS environment allowed them to scale their computational fluid dynamics environment and reduce simulation time by over 80%.[3] The faster they can learn, the more successful they can be.

Architect for Reusability

A top concern of nearly every organization today is speed and value for money. The easiest way to go faster and be more cost-effective is through reuse. Investing in architecting for reuse requires modularity and standardization with loose coupling between components and patterns of abstraction. In addition, each of the components must have disciplined version control with the ability to support backward compatibility.

Modularity is the degree to which the components can be separated and recombined. For example, we can change an embedded system with a field-programmable gate array (FPGA) by simply reprogramming the FPGA to implement a different function, whereas an embedded system containing a fully integrated circuit would require a whole new circuit.

Standardization, especially for interfaces, enables a system to rapidly change by providing the ability to plug in a new component as long as it meets the interface standard. For example, CubeSat Serial Interface, developed by NASA, allows actuators to be easily replaced.

Architecting for reuse requires loose coupling between components, where each component in the system has minimal dependencies on another. One example of loose coupling in satellite systems can be seen in attitude control systems: a change to a sensor should not impact anything associated with the magnetometer. The last key to reuse is leveraging patterns of abstraction, which is borrowed from software design patterns where complex implementation details of the system are hidden behind simplified interfaces. For decades, we have known that rockets were one-and-done until SpaceX decided to change the game with reusable launch technology. SpaceX reduced launch costs by roughly 60%, and as of December 2021, they have had one hundred landings.[4] Key attributes that allow us to increase reuse are standardization, modularity, and patterns of abstraction.

Architect for Security

Security has grown in importance year over year. As we have moved to open architectures for improved performance, the attack surfaces of our systems have exploded. We can ask our digital friend Siri to turn on the patio lights, adjust the temperature in the house, and tell us how many loads of laundry were done this week. This level of convenience has been obtained through the Internet of Things (IoT), the creation of a collective network of smart devices that communicate between the cloud, as well as themselves. The drawback of this level of connectivity is that the number of possible points where an unauthorized actor can access our systems has grown exponentially.

This problem becomes more untenable when we talk about cyber-physical systems, where we are integrating sensing, computation, control, and networking into physical objects and connecting them to the greater IoT. Security is a necessary feature of the cyber-physical systems architecture to ensure that capabilities are not compromised by malicious agents—and that the information used, processed, stored, and transferred has its integrity preserved and is kept confidential where needed.

The nature of cyber-physical systems not only increases the consequences of a breach but also introduces additional types of vulnerabilities. For example, timing in a cyber-physical system has unique vulnerabilities. Security needs to be built into cyber-physical systems by design to be sufficiently flexible to support a diverse set of applications. Thus, the focus of developing an intentional security architecture is the only way to decrease the risk of nefarious access to our systems. A high-level example of security architecture for our CubeSats is illustrated in Figure 7.2.

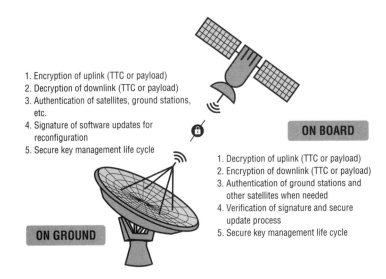

1. Encryption of uplink (TTC or payload)
2. Decryption of downlink (TTC or payload)
3. Authentication of satellites, ground stations, etc.
4. Signature of software updates for reconfiguration
5. Secure key management life cycle

ON BOARD

1. Decryption of uplink (TTC or payload)
2. Encryption of downlink (TTC or payload)
3. Authentication of ground stations and other satellites when needed
4. Verification of signature and secure update process
5. Secure key management life cycle

ON GROUND

Figure 7.2: Security Architecture for CubeSat

There are a couple of well-known patterns that support security in systems, including Defense in Depth (DiD) and Zero Trust:

- **Defense in Depth:** DiD uses a layered approach of defensive mechanisms to protect systems down to the data layer. There is extensive redundancy ensuring that if one line of defense fails, the next one can pick it up. Think of *Mission Impossible,* where Tom Cruise's character must overcome multiple layers of defense to obtain the list in the CIA headquarters. A layered DiD example is illustrated in Figure 7.3.

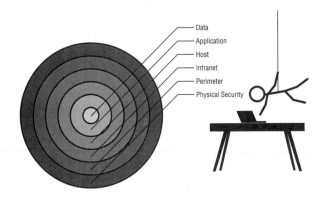

Data
Application
Host
Intranet
Perimeter
Physical Security

Figure 7.3: Defense in Depth

- **Zero Trust:** Often implemented with Defense in Depth, Zero Trust is a principle of least privilege. Basically, the system eliminates implicit trust for all actors in the system. The trust is explicitly provided at each layer. For example, just because an individual was able to enter the Cineplex does not mean they can enter a specific movie theater. They are granted entry only after sharing their ticket for a specific movie. The three tenets of Zero Trust include risk awareness, least privileged access, and continuous access verification. A simple Zero-Trust architecture containing policy-based access control (PBAC), attribute-based access control (ABAC), and role-based access control (RBAC) is illustrated in Figure 7.4.

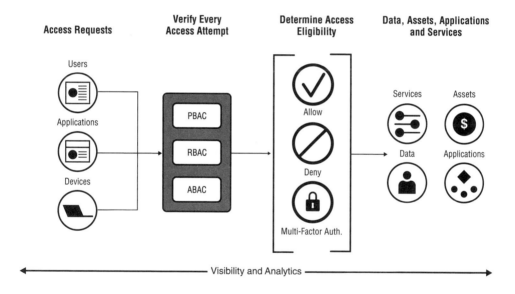

Figure 7.4: Zero-Trust Architecture

Architect for Manufacturability

When first designing a system, we rarely think about how that system will be assembled (i.e., the manufacturing). However, as cyber-physical systems have several cost and schedule impacts, it's worth pulling this discussion to the beginning of design. We can increase speed and reduce time in manufacturing by standardizing common components and creating processes that can be easily replicated.

The next step is to simplify the process by minimizing the number of parts. Key attributes of manufacturability include simplicity, ease of assembly, minimized part count, and cost-effective materials. There is a trade-off to be made for each architectural consideration, depending on our business objectives.

IKEA furniture is well known for their easily assembled furniture. I (Robin) purchased a bookcase in 2015, where I received a flat-packed box with predrilled holes, dowels, and screws that I was able to put together myself, which was a completely different experience than I had years before with the Barbie Dreamhouse. One of the items was built for ease of assembly and one was clearly not.

Tesla builds their battery pack as the structural platform in one big unit. They have traded off the ability to make rapid change for manufacturability; the new design reduces the number of parts and the total mass of the battery pack and improves the range and efficiency of the vehicle electronics.[5] AK-47 guns are designed with cost-effective materials that can be easily modified. Currently, Ukraine is using 3D printers on the battlefield to enhance AK-47s for real-time support in their effort to protect themselves against their invaders.[6]

Relativity Space radically reduced the parts for the Terran R, a fully 3D-printed reusable rocket that is scheduled to launch in 2024.[7] All of the use cases we have discussed here prioritized manufacturability as a key enabler to deliver quality at speed.

Architect for Agility

Architecting for agility allows a system to adjust and adapt rapidly through modularity and standardization. The result is the ability to deliver higher-quality products and services to market faster. There are multiple considerations when we discuss architecting for agility, which include configurability, deployability, and maintainability.

Architect for Configurability

Architecting for configurability involves designing a system that can be easily and efficiently configured to meet different requirements and use cases. Configurability is a key consideration in Industrial DevOps. Technologies such as microservices, containerization, and automation have allowed companies building everything from automobiles, space vehicles, and manufacturing plants to create software-defined systems. Software lives forever, which means it's the ultimate tool in configurability.

FPGAs are integrated circuits that can be programmed and reprogrammed after they have been manufactured, which enables space systems to dynamically change guidance, navigation, and control systems for satellites to adapt to changing conditions. Making use of design patterns such as modularity, parametrization, configuration files, and feature toggles supports building systems for configurability.

Two attributes to consider when architecting your system for configurability are compatibility and testability. A system that is designed to be compatible with a wide range of sensors, communication protocols, and other hardware components can be easily adapted to different use cases and environments. Instrumenting your system to be easily testable is also important for cyber-physical systems, because they often have several regulatory requirements. If each configuration change results in a long, manual validation, then we are not really getting the benefits of configurability.

Advancements in Product Line Engineering (PLE) and DevOps have greatly impacted how we architect for agility. PLE is the engineering of a product line using shared engineering assets-based commonalities across the products and the planned variations. It's similar to the patterns of abstraction we discussed in architecting for reuse, where we used simplified interfaces to let us enhance systems.

Lockheed Martin's LM 400 is an excellent example of this approach. They have created a common LM 400 bus utilizing Modular Open Systems Approach (MOSA) standards for interoperability. The bus can be configured to scale up with higher power or larger payloads to execute a variety of missions. The LM 400 also utilizes Lockheed Martin SmartSat technology, which allows you to adapt the mission capability by uploading an app from the ground station and reconfiguring the satellite.[8]

Architect for Deployability

One element that allows us to deliver faster is to make it very easy to deploy changes to the system. Areas we need to consider are packaging, installation, and automation. Attributes to consider when building systems for deployability include installability, upgradability, and portability. Advancements in areas such as infrastructure-as-a-service (IaaS), platform-as-a-service (PaaS), and software-as-a-service (SaaS) (see Figure 7.5) have reduced labor and time in installing and upgrading.

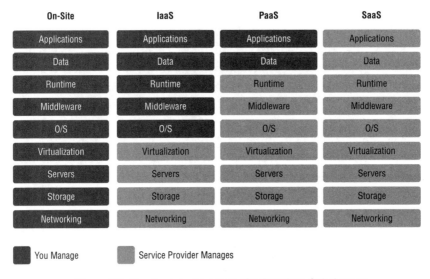

On-Site	IaaS	PaaS	SaaS
Applications	Applications	Applications	Applications
Data	Data	Data	Data
Runtime	Runtime	Runtime	Runtime
Middleware	Middleware	Middleware	Middleware
O/S	O/S	O/S	O/S
Virtualization	Virtualization	Virtualization	Virtualization
Servers	Servers	Servers	Servers
Storage	Storage	Storage	Storage
Networking	Networking	Networking	Networking

■ You Manage   ■ Service Provider Manages

Figure 7.5: The Journey from Fully On-Premises Solutions to
Fully Off-Premises Solutions

- **Infrastructure-as-a-service** is a cloud-computing model where the networking and storage are managed for us, and teams need to worry only about building the capability from the operating system up.
- **Platform-as-a-service** is a cloud model that manages everything up to the data and applications. This allows teams to focus on rapidly building applications and deploying on an existing platform.
- **Software-as-a-service** is a cloud model that manages everything for the user. The only thing teams need to concern themselves with is subscriptions to applications for use such as email.

The technologies allow for the delivery of much faster updates to cyber-physical systems. In addition to the cloud-compute models, containerization has greatly increased portability by bundling portable self-contained units known as containers that can be deployed in any compatible environment. Containerization has become a de facto standard in the automotive industry, where companies like Volkswagen are focused on building software-defined vehicles capable of taking over-the-air updates. [9]

From an operations perspective, technologies mostly affect cyber-physical systems from the peripherals, such as areas where we want to track, process, and analyze large amounts of data. For example, areas such as satellite communications and data downlink are heavily leveraging these technologies. Cloud services in the satellite domain are predicting around $16 billion in revenue for the 2020s.[10]

Architect for Maintainability

We can build great systems faster, but if they are not easily maintainable, their value is quickly reduced over time. Our experience in aerospace and defense has showcased this problem across most legacy systems. Many of the legacy systems are monolithic applications that are very difficult to maintain. Digital modernization is almost code for refactoring out monolithic systems.

Attributes to consider for maintainability include testability and ease of development. The ease of development is one of the biggest elements we consider for maintainability. Currently, the Toyota Prius has been credited with being the car with the lowest maintenance costs over ten years. The reasons for the low maintenance costs of the Prius are attributed to high-quality standardized components and simplified systems designed for easy access.[11] Your car is designed for maintainability. We know that our car tells us about everything from low tire pressure to the need for an oil change. The automotive builders architected their cars to capture a wide range of data and use that data to make life easier for their owners.

Technology Trends Impacting Systems Architecture

Architecture has always been important, but in the last few years, there have been some exciting new tools for our toolbox that can support architecting cyber-physical systems, including cloud computing, microservice architectures, DevOps, artificial intelligence, and edge computing. While we've spoken about some of these in previous sections, let's take a moment to look at these with a bit more depth, as they are keys to our ability to architect for speed.

Cloud Computing

Cloud computing allows us to use shared resources for our individual infrastructure needs. The benefits of cloud computing for cyber-physical systems include flexible scaling, reliable connectivity, large-scale data storage and processing, extensive collaboration, and reduced cost.

One of the largest architectural considerations for cyber-physical systems is latency. Real-time embedded systems have significant requirements for rapid deterministic response times and cannot handle latency in the system. To overcome the risk, we must focus on network performance and leverage edge-compute options where possible.

GE Aviation leverages a cloud-based platform called Predix that enables airlines to collect and analyze data from sensors and devices on aircraft. This allows them to improve maintenance, reduce fuel consumption, and enhance the passenger experience by collecting data from engine sensors, flight recorders, and weather data. While GE has not publicly disclosed the amount, they have confirmed cost savings in fuel and maintenance from their cloud-computing platform.[12]

Microservices

Over the years, architectures have been migrating from monolithic to microservices and now are even trending toward serverless, as illustrated in Figure 7.6. Microservices are a software development architecture pattern that involves building small, independent services that work together to form a larger application. Each microservice is designed, developed, and deployed independently. The benefits are exponential because we can reduce dependencies in both the system and the organization.

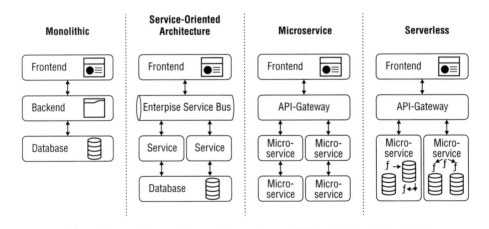

Figure 7.6: The Evolution of Architectures

Digital natives like Amazon are known for this type of architecture and their ability to deploy new updates to their systems multiple times a day.

However, while microservices help us go faster and improve scaling, they also add increased complexity and integration challenges.

While there are challenges such as heterogeneity, microservices can be used in cyber-physical systems with some adaptations. There is still more work to be done in this area, but it seems that Tesla has cracked the nut. The Model S has multiple systems made up of microservices, but perhaps the most famous is Tesla's operating system itself, which allows Tesla to perform over-the-air vehicle updates.[13]

DevOps

DevOps is one of the newer methods that continues to mature, enabling us to move to a place of what Isaac Sacolick referred to as continuous architecture or the "move from a former waterfall approach, where architecture was done mainly before features were implemented, to a continuous runway."[14] Principles include architecting "long-term products, not just projects solutions" and "validating the architecture by implementing."[15]

In cyber-physical systems, the challenge in implementing comes from hardware-embedded systems, which are tightly integrated hardware components, and limited resources, such as processing power, memory, and storage. While digital natives like SpaceX regularly make use of DevOps, more traditional companies are also making use of DevOps for large cyber-physical systems. The Air Force and Northrop Grumman leveraged what was referred to as DevStar, a play on DevOps, that focused on speed, quality, focus, and collaboration for the new B-21 Raider that was revealed on December 2, 2022.[16]

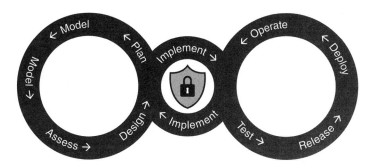

Figure 7.7: TwinOps

Another example of DevOps in the cyber-physical world is TwinOps (Figure 7.7), the practice of using digital twins. As we discuss more in the chapter associated with testing, digital twins are making things possible for hardware that were not previously. By creating a digital twin, developers and operators can simulate and test changes to the system in a safe, controlled environment before implementing them in the real world.

Artificial Intelligence (AI) and Machine Learning (ML)

Artificial intelligence (AI) is the simulation of human intelligence in machines that are programmed to perform tasks such as learning, problem-solving, decision-making, and perception. Cyber-physical systems are not immune (see Figure 7.8). We all know vehicle manufacturers are in a race to create autonomous driving, but they leverage AI, even in factory operations. BMW developed software that creates AI applications for object recognition, enabling them to label objects in images offline, to quickly create AI apps that can then reliably identify those objects in images during production.[17] Machine learning (ML) is advantageous because of its ability to address a variety of special areas to improve performance and quality.

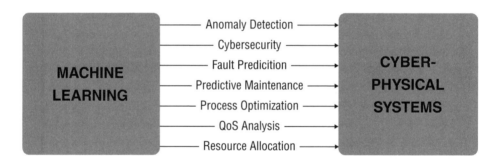

Figure 7.8: AI/ML in Cyber-Physical Systems

The benefits of this technology range from efficiency to accuracy, speed, and more. Using AI/ML, architects and systems engineers can evaluate multiple design options with multiple parameters to remove bad options quickly and amplify engineering knowledge. The question is not, "When should we use artificial intelligence?" It is "In what scenario would we not use artificial intelligence?"

Edge Computing

Edge computing moves computer storage and processing to the edge of the network, where it is closest to users, devices, and data sources (see

Figure 7.9). When data is processed at the edge, the latency and bandwidth requirements of the network are reduced for faster processing of data. Edge computing is speeding up everything from anomaly detection to space exploration. The Hewlett Packard Enterprise (HPE) Spaceborne Computer-2 (SBC-2) onboard the International Space Station (ISS) and a custom edge solution created and managed by IBM have proven that they can reduce the time to analyze large amounts of data from weeks to hours, which is critical in future space exploration because of the need to rapidly obtain data on astronaut health, spacecraft integrity, and the condition of plants being grown as a food source.[18] Architectural considerations for edge computing include rugged hardware, sufficient storage, and rich connectivity.

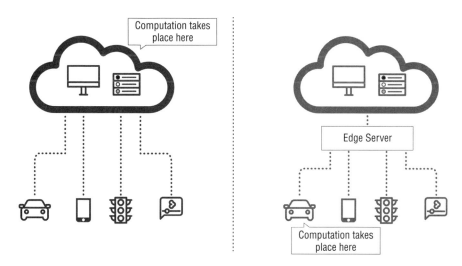

Figure 7.9: Cloud Computing vs. Edge Computing

Digital Twin

Digital twin architecture involves creating a virtual model of a physical system, enabling simulation, monitoring, and analysis of the system's behavior in real time. This architecture is being used in industries such as manufacturing, health care, and aerospace. Digital twins can also be used to support system design and development by allowing designers to prototype and test new systems in a virtual environment before deploying them in the real world. This can help to reduce development costs and accelerate time to market, as well as identify potential issues before they become major problems.

Digital twin is a concept that moves the physical system into cyber-space. It is not one product or technology but rather the amalgamation of multiple technologies that include simulators, emulators, big data, IoT, 5G, and AI. Digital twin is a digital representation of the physical system in cyberspace. Digital twins are much more than simulators, because they are connected to the physical system where we can get ongoing telemetry to provide increased fidelity of our tests. This brings together a host of technologies to the table such as modeling, simulation, emulation, digital twinning, augmented reality (AR), virtual reality (VR), and additive manufacturing.

Simulator

A simulator is a machine or program that simulates a real-life scenario for the purposes of being able to experiment with many different options. Here are some of the various types of simulations:

- **Emulator:** This is hardware or software that enables one computer system (called the host) to behave like another computer system (called the guest). An emulator typically enables the host system to run software or use peripheral devices designed for the guest system.
- **Digital Ghost:** This is a new technology from GE that makes use of an avatar to detect anomalous behavior. The Digital Ghost localizes where the behavior is occurring, such as on a sensor, actuator, or control node. It then provides real-time insight into tampering and ensures minimal degradation or a graceful shutdown, defending critical infrastructure from cyberattacks.
- **Digital Twin Prototype (DTP):** The DTP is a prototype of a physical asset that contains the recipe, like the design pattern discussion; this would be similar to an abstract factory pattern.
- **Digital Twin Instance (DTI):** Created from a DTP, this twin has been customized to meet a specific physical asset, where it is linked, passing back and forth real-time data from sensors and aggregators.
- **Digital Twin Aggregate (DTA):** The DTA is an aggregate of multiple DTIs; the DTA uses the data from many instances for even more extensive data (wisdom of crowds).
- **Digital Twin Environment (DTE):** This is a collection of DTP, DTI, and DTA in one ecosystem to enable predictive operations.

- **Digital Thread:** This is the use of digital tools to signify the digitalization and traceability of a product throughout the life cycle (see Figure 7.10).

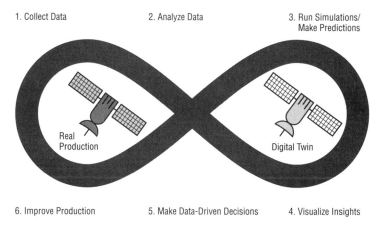

1. Collect Data   2. Analyze Data   3. Run Simulations/
                                                    Make Predictions

Real
Production   Digital Twin

6. Improve Production   5. Make Data-Driven Decisions   4. Visualize Insights

Figure 7.10: Digital Thread

Accelerators for Architecture

As we have discussed, there are a lot of elements that impact architectural design. However, by and large, four areas that enable you to adapt your architecture for speed are modularity, standardized interfaces, iterative flow, and broad collaboration. Modularity allows you to plug and play new capabilities like LEGO® blocks, standardized interfaces provide a known communication mechanism between modules, iterating for flow is the balance between emergent and intentional design, and broad collaboration minimizes bottlenecks and creates transparency into the system. These elements are common in systems that we have seen are evolving rapidly, such as the case studies highlighted at the close of this chapter.

Modularity

As mentioned before, organizational structure changes in concert with the architecture. A flat structure organized around value requires a loosely coupled modular architecture. However, loose coupling is more than a simple measure of the count of services in a system. To build a modular architecture, we must begin with the system's responsibilities and its components. Each component should have a clear responsibility. The next step is to define interfaces, how the modules communicate with one another.

Each module should be self-contained and developed, tested, and deployed independently. This approach allows developers to build larger and more complex applications by assembling independent modules. Modular architecture is highly scalable and extensible, making it ideal for large and complex applications.

MOSA is a framework for developing complex systems, particularly in the defense industry, and can work well in cyber-physical systems. MOSA is a structured approach to building in modularity, open standards, and interoperability. The idea is to develop systems that are composed of interchangeable, standardized modules or components that can be combined in different configurations to meet different requirements. This allows for greater flexibility and adaptability, as well as easier maintenance, upgrades, and repair. You can think of this in terms of LEGO® building blocks that help us compose systems. A good example of a modular architecture is illustrated in Figure 7.11

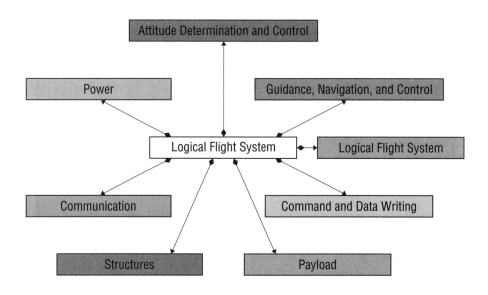

Figure 7.11: Modular Architecture Example

Standardized Interfaces

A standard interface is a set of rules or protocols to interface at the boundaries. In hardware, it can be a plug, socket, or cable. In software, it's typically an application programming interface (API). Modular architectures and standard interfaces are a must to scale. For cyber-physical systems, we need to incorporate a variety of interfaces.

If we look at our CubeSat example, standardized interfaces enable CubeSats to be quickly and easily integrated into larger systems, such as launch vehicles or other satellites. The CubeSat Kit is a standardized set of components and subsystems that can be used to build a CubeSat. It includes standard dimensions for the satellite, as well as common interfaces for power, data, and communication.

The PC104 standard is a widely used form factor for embedded computing systems, and it is commonly used in CubeSat design. The PC104 interface provides a standard way to connect and communicate with different components of the satellite. The CubeSat Space Protocol is a communication protocol that is designed specifically for CubeSats. It provides a standard way for CubeSats to communicate with each other and with ground stations. The CubeSat Deployer is a standardized mechanism for deploying CubeSats from a larger spacecraft, such as a launch vehicle or another satellite. The deployer provides a standard interface for attaching and releasing CubeSats, making it easier to integrate them into larger systems.

Iterative and Incremental for Flow

In cyber-physical systems, it is important to balance intentional design with emergent design; resist the urge to do big up-front design. The goal is to identify what the system needs right now and then create enablers and spikes to incrementally build out the architecture. Begin with the most basic system you can learn from and incrementally add capabilities by defining new features you want added, then refactoring the architecture to meet the needs of the features. This is sometimes called iterative and incremental design.

This goes hand in hand with modular design and is an enabler for incrementally building out the architecture. Each module is independent, with a discrete set of responsibilities that allows us to test the system by mocking out other components that have not been completed. Modules of capabilities can then be validated.

The Mars Rover was initially designed to navigate preprogrammed routes. Over time, its architecture evolved, giving it the ability to detect and avoid objects, making decisions based on the environment.[19]

Collaborative Team Sport

As we have discussed, architecture enables us to design, communicate, and maintain systems, which requires it to be a team sport. We have all heard of the ivory-tower architect who makes exquisite architectures while being isolated from the reality of the system. The problem is that the smartest person in the room is not as smart as the team. The architect is a member of, not separate from, the team.

In large legacy cyber-physical systems, the system has often drifted from the original designs and models, and the architects are not aware of the current configuration of the system. This causes them to make decisions without understanding the impacts to the current state. While the ivory-tower architect may succeed in design, this pattern fails to meet the need to communicate or maintain the system.

Another problem can occur when collaboration *within* a team is strong but collaboration *across* teams is not. A team of modelers has designed an amazing set of detailed models for our system of interest. But what if the users cannot read SysML? In a recent presentation given by an executive in the Air Force, we were told they were confident that the vendors could provide the models needed, but they were not confident that the airmen would understand the models provided, which means they really could not provide feedback. Again, we had succeeded in design and maybe increased our ability to maintain the system, but we had still failed to communicate the design to stakeholders. This would be similar to someone giving you a book written in Mandarin and asking your thoughts on the story—except you don't speak Mandarin. Without an interpreter, we have nothing to provide. For these reasons, we need a collaborative team architecting easy-to-build and easy-to-read blueprints to accelerate systems architecture.

Conclusion

In this chapter, we addressed the misnomer that Agile or DevOps teams do not need intentional architecture. We have discussed the need for visual tools such as a checklist to perform trade-off analysis of the architectural considerations. Optimizing for one consideration is most likely going to cause us to suboptimize for another. We have explored all of the architectural considerations to provide an understanding of how they affect cyber-physical systems. We described the importance of easy-to-build and easy-to-read blueprints for our architecture. If the blueprints are not easy to build, they will not be maintained. If the blueprints are not easy to read, they will not be understood by our stakeholders. We analyzed a wide variety of new technologies such as cloud computing and AI, which are now impacting the architectural trade-offs being made. Lastly, we have provided a set of accelerators to support architecture.

CASE STUDY: Planet Labs: The Agile for Aerospace Approach

Planet Labs is an American-based private imaging company that was founded in 2010 with a mission to image all of Earth daily to identify temporal global changes. The imaging data allows them the ability to analyze agricultural, energy, forestry, maritime, and sustainability events and impacts. There are a number of companies that perform Earth imaging, such as DigitalGlobe and Maxar Technologies, but Planet Labs is unique in that they refer to their approach to delivery as Agile Aerospace, which they describe as a philosophy of spacecraft development through rapid iterations.

Planet Labs is what we consider a very early adopter of Agile and later DevOps practices at the system level. They reportedly began their approach in 2012 and since then have completed fourteen major iterations of the Dove spacecraft design, which is even more than is reported from SpaceX. The key difference that has to be noted is that SpaceX has to focus on high-performance and safety-critical features, whereas Planet Labs' driving goals are cost-effectiveness and the sustainability of satellites.

Planet Labs focuses on optimizing spacecraft architecture to enable rapid evolution of the Dove spacecraft design by using some of the accelerators we discussed, such as modularity, standardized interfaces, and open architecture. In addition, they use commoditized components that can be easily purchased and replaced. They create redundancy in their system by leveraging an entire constellation where if one CubeSat fails, they are still able to deliver a stream of high-quality imagery. Planet Labs invests heavily in developing intelligent constellation management software, which has the benefit of being able to detect and respond to satellite problems and reposition to ensure there is not a gap in coverage if one fails.[20]

The results of Planet Labs' approach are that they were able to get satellites in space in record time, enabling them to learn quickly about their entire value stream, including regulatory, launch, and integration challenges, before it became too expensive to make changes to their design. Their architectural approach has allowed them to continuously optimize their design. They report they build over forty Dove satellites a week and make use of a ride share agreement with SpaceX that has enabled them to launch over five hundred satellites to date. According to

Forbes magazine, Planet Labs is considered to be five to seven years ahead of any competition.[21]

Joby Aviation is an American aerospace company that was founded in 2009 by JoeBen Bevirt, a technology entrepreneur. Joby has developed an electric vertical takeoff and landing aircraft for urban air mobility (UAM) applications.[22] Joby plans to launch an air-taxi service in 2024 where they plan to bring air mobility to the masses. Joby has some very innovative big-name investors in their vision, including Delta, Uber, and Toyota. In addition, Joby has contracts with the DoD to support use cases such as relocation of personnel and medical response.[23] The UAM market is expected to grow rapidly due to the need for efficient and sustainable transportation in urban areas. Key competitors in this space include Airbus, Volocopter, and Hyundai Motor Company, to name a few.

Joby projects to have 141 aircraft in operation and be generating revenue by the end of 2024, with a forecast of nine hundred aircraft in the air by 2026.[24]

Joby Aviation has a modular architecture with standardized interfaces and a complete delivery pipeline that allows them to rapidly iterate on changes to the vehicle. They make use of test-driven development across development of the entire vehicle to ensure they build quality in.

GETTING STARTED

- Create a checklist of nonfunctional considerations to complete a trade-off analysis against vision.
- Build an easy-to-follow blueprint to design, communicate, and maintain.
- Utilize MOSA architectures in design.
- Invest in standardized interfaces.
- Design for flow.
- Collaborate across all stakeholders.

KEY TAKEAWAYS

- Modular architecture is a foundation for successful Industrial DevOps implementation.
- Architecting a system requires making architectural trade-offs.

- We need easy-to-build and easy-to-read blueprints for architecture to be effective.
- The architecture evolves as the system evolves.
- New technology trends have impacted how we architect for systems.
- Modularity, standardized interfaces, designing for flow, and collaboration accelerate architecture.

QUESTIONS FOR YOUR TEAM TO ANSWER

- What are the highest-priority objectives of your system?
- Which technologies are impacting your architecture design?
- Which accelerators have you selected for your architecture?
- What trade-offs in design did you consider? Why?
- What level of fidelity do your models need to be for you to design, communicate, and maintain your system?
- Can your stakeholders understand the models? Do they need an interpreter.

COACHING TIPS

- Remind your teams why architecture is important for Industrial DevOps.
- Use a visual tool, such as a checklist, to prioritize trade-offs between the nonfunctional requirements.
- Begin with the constraints such as compliance, security, and safety; architecting these into the system after is ineffective and creates extensive rework.
- Use right-sized models and artifacts. If they can't be maintained or read, they are shelfware.
- Bidirectional traceability is necessary to continuously verify and validate the system.

CHAPTER 8

ITERATE, MANAGE QUEUES, CREATE FLOW

> **PRINCIPLE 5:** Iterate, manage queues, and create flow to emphasize the importance of fast feedback, experimentation, continuous learning, and managing queues to improve flow.

All I Really Need to Know I Learned in Kindergarten. This book, published in 1986, is a set of short essays on the basic rules we learned in kindergarten, mostly about life and getting along with others. And what else do we know about kindergartners? They experiment. They learn through repetition and iterations of experimentation.

This was demonstrated and made popular through the famous marshmallow tower challenge. You may even have participated in this activity at some point, as it's often delivered as a team-building exercise. But there is more to it than that. First, if you are not familiar with the activity, it takes eighteen minutes. In that time, each small team can use twenty sticks of spaghetti, one yard of tape, one yard of string, and one large marshmallow to build the tallest structure with the marshmallow on the top.

This experiment has been done thousands of times across all age groups, from schools to the corporate world, and across functional areas. What has been discovered is that kindergarten students beat the average. Why? They talk a little and then jump in and get going. They see what works and doesn't work, then they quickly iterate and improve the structure.

On the other hand, business graduates tend to do a lot of talking and planning, then struggle at the end when their plan doesn't work and they have no time left to figure out what went wrong. In the business context, their schedule is late, even if their plan was great.

A similar lesson was learned with the Wright brothers at Kitty Hawk, North Carolina, versus the team led by Samuel Langley, secretary of the Smithsonian Institution. Both had been working tirelessly to achieve the first flight. The Wright brothers reached that goal first. Why? They experimented, tested, and used learning to make the next iteration of improvements. Their natural curiosity and desire to learn helped achieve the desired outcome. Not only did they achieve time to market faster, but they did so at lower cost. It wasn't their education (they were not STEM grads), and it wasn't money (they spent less than $1,000, while the Langley team spent about $70,000*).[1] Orville Wright was once asked what the difference was between his team and Langley's, and he responded, "The greatest thing in our favor was growing up in a family where there was always much encouragement to intellectual curiosity."[2]

You can still see these differences today. Typically, at the start of a new project, a project manager, working with their team, creates detailed plans and a predictive schedule that lays everything out over the course of a couple years. This plan shows how all the parts come together. Unfortunately, these plans are created at the point in the product development cycle where less is known about the solution. Even if the team did know everything, once the project is started, it will soon change based on new learnings, thus creating a need for reprioritization. We aren't implying that planning is not important. It *is* important. And what is more important is the right level of planning that enables teams to inspect, adapt, and improve processes. It is about the learning as they drive toward a desired outcome.

We have discussed the importance of multiple horizons of planning (Chapter 5). Plans will evolve and new understanding will be gained. The ability to respond to change is paramount for success. The iterative nature of Agile and DevOps coupled with digital capabilities accelerates learning and provides fast feedback for improved value demonstration. Just as the Wright brothers and the marshmallow challenge demonstrated.

According to the *Harvard Business Review* report *Competitive Advantage through DevOps*, the implementation of Lean and Agile practices and their iterative nature yields an advantage of innovation. Of 654 companies responding, they confirm that Lean, Agile, and DevOps are positively impacting their business: increases of 70% speed to market, 67% productivity, 67% customer relevance, and *66% impact on innovation*.[3] (See Figure 8.1 for an illustration.) Agile teams in the cyber-physical space can leverage the iterative process to build a culture of innovation and experimentation.

In this chapter, you will learn the importance of iterating with fast feedback and continuous learning, managing queues, and improving flow. These practices enable teams to explore and validate systems regularly,

---

* Today, that would be < $28,000 and ~$2 million, respectively.

confirming the team's understanding of the system they are building as they strive to build better systems faster.

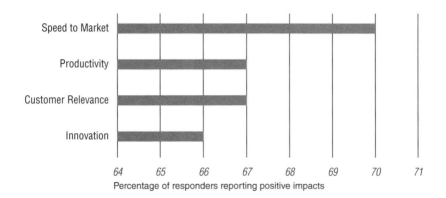

Percentage of responders reporting positive impacts

Figure 8.1: Positive Impact of DevOps

Source: Harvard Business Review Analytic Services, Competitive Advantage through DevOps*.*

Iterate

Industrial DevOps invites teams to experiment and continuously learn and improve. With each iteration of development, we gain feedback from stakeholders and the solution evolves. Slowly, the way forward becomes clearer. Agile teams iterate along a path, starting from uncertainty and driving toward a defined solution as they approach their release (Figure 8.2).

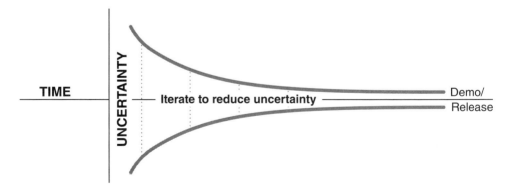

Figure 8.2: Iterate to Reduce Uncertainty

Learning gained through iterative development is tightly connected to the Industrial DevOps principle of implementing data-driven decisions. Together, the two principles create an environment in which teams iterate toward a solution by providing demonstrable results and objective evidence that is used for improved decision-making. Due to the tight coupling of these two principles, you will see some overlap in how we describe them. Iterating and getting feedback tied with data-driven decision-making enables stakeholders and customers to make better, more informed decisions on the next-best step to take in delivering solutions that meet user needs and achieve business outcomes.

As early as 1945, Herbert Simon's research showed that management will never have all the information needed to make a perfect decision; instead, they seek good-enough decisions based on the data and information at hand.[4] In 1987, his research on problem-solving and decision-making added the importance of learning "for successful adaptation to an environment that is changing rapidly."[5]

Nearly forty years later, we are still addressing this challenge. However, now we have the power of advanced digital capabilities and tools to improve the visualization of progress across the value stream. We can also observe integrated demonstrations of functionality performed by teams. Together, this creates a more sophisticated and improved decision-making process for managers and stakeholders.

The iterative development cycle is similar to the plan-do-check-act (PDCA) cycle popularized by W. Edwards Deming and the Lean community. With the PDCA cycle, you plan (P) a little for the next change, you build and test (D) the change, you analyze the metrics and results (C), and then, based on the learning, you decide how to act (A) next. With our iterative development cycle, each iteration is typically one to four weeks and traverses through similar steps based on the stories prioritized for that iteration. The results are demonstrated and analyzed throughout the iteration. Based on the feedback from the stakeholders and what the team learns from the demonstration, they plan their next steps. It is about continuously improving and assessing progress toward the target. Learn from the lessons of the Wright brothers: iterate, refine, and improve toward a defined goal.

Through experimentation and feedback, the team continues to make progress toward the goal—that is, the value delivered to the customer. With this process, "New products, more or less, are allowed to emerge from the collective learning at all levels of the system."[6] Therefore, Industrial DevOps employs the concept of set-based design, an approach described by Michael Kennedy in his book *Product Development for the Lean Enterprise: Why Toyota's System Is Four Times More Productive and How You Can Implement It*. Set-based design with iterative development cycles provides the ability to "explore multiple sets of possibilities at the subsystem level against broad targets"[7] and proactively explore the limits of hardware design.

This is a different approach to development than point-based design, in which teams build to a predefined set of specifications. These specs are often very detailed and leave little opportunity to explore options. With iterative or set-based design, development teams experiment and gain knowledge regarding different design options. They are then able to evaluate multiple possibilities before making costly physical component decisions.

As teams build their cyber-physical system using Industrial DevOps, they embrace these practices. For example, let's look at a team defining a guidance, navigation, and control system. First, they will iterate and test out the algorithm to determine the attitude error so that it may adjust the attitude controller. The team uses both simulated and emulated environments to test and see how changes are working. They then engage across all functions and integrate small batches from development to manufacturing for regular testing and feedback.

At the end of each iteration, the team validates whether they're going in the right direction and how well the newly integrated components are working. This is the concept of "design a little, communicate a little, build a little, verify a little, and iterate" introduced in the principle of architecting for speed (see Chapter 7).

Only through a loosely coupled architecture are teams able to work independently enough to develop, test, and deploy iteratively with little coordination overhead. With tightly coupled systems, one team could make a change to the system, and it would have a rippling effect across several teams' development, causing unintentional chaos.

When our teams discover a solution is not working as anticipated, they are able to quickly pivot and explore and test other options. These small checkpoints are like taking a long road trip and periodically checking that you are on the right course, that you have not deviated from the intended direction as a result of travel impediments along the way.

Manage Queues and Improve Flow

Imagine a local company has three hundred employees who head to the cafeteria at noon for lunch. The capacity of the cafeteria was not designed for that many workers at the same time. The result is long lines and long wait times. Grumbling starts to happen, and soon there are discussions on the need to expand.

While expansion may be a viable option, another option might be to stagger lunch breaks so that only a portion of the workers enter at 11:30, 12:00, and 12:30. Once lunch breaks are staggered, fewer people enter the cafeteria at the same time. This results in less wait time, a steady flow of people going through the cafeteria system, and happier workers.

The analogy holds true for the products we develop as well. The more features we have in progress, the slower the feature delivery. Working fewer features at a time reduces wait time and queue length. As we learn from Reinertsen, "it is not that queues themselves are bad, it is the economic loss associated with them."[8]

For example, let's look at one of my (Suzette's) experiences at a small startup during the dot-com surge. The company was a hosted application provider, and I was a product delivery manager. Only once features were delivered did we start generating revenue; up until that point, it was mostly all cost to the company. Striving to meet customer needs meant defining a minimum viable set of features and delivering them as quickly as possible. Through my team's own experimentation, we learned how to limit WIP to improve the flow of work and deliver features more quickly. My team delivered on time. While my team and I had not been formally trained on these practices, we understood the need for improving flow to deliver to the customer at the speed of relevance.

Queuing Theory

The concept of flow and throughput is based on the concept of *queuing theory*. The importance of queuing theory is based on early work by A. K. Erlang in 1904 to help determine the capacity requirements of the Danish telephone system as they were trying to determine the number of telephone lines needed based on usage.[9] Since that time, people have been studying queues and how to optimize throughput, and the idea was made popular by the Lean manufacturing community.

While queuing theory has been studied for many years, its application is not always well understood, nor are the impacts it has in system development and delivery. To help understand the concept, we have identified several key phrases and definitions, which are defined in Table 8.1.

Table 8.1: Queuing Theory Concepts

Term	Definition	Example
Queuing theory	A mathematical study of delays in waiting in line	Tasks queuing up in a computer system
Queuing system	Multiple interconnected queues	Manufacturing systems
Queue	A line of things waiting for processing	A line of people waiting at Starbucks for beverages
WIP	A partially finished product or service awaiting completion.	An individual working multiple activities at one time
Throughput	Average processing rate of the queue	Average number of people serviced at the DMV per hour
Theory of Constraints	Identifies the limiting factor in throughput, frequently referred to as *bottleneck*	People, subject matter experts, machine capacity

Kanban

To help manage queues and the flow of work, many teams in development and manufacturing have used a tool called kanban, which was originally started in Lean manufacturing. The idea originated when Toyota visited grocery stores in the United States and observed a just-in-time flow process. As customers pulled supplies from the shelf, it would trigger the process behind it to refill the shelves. Supplies were delivered (restocked) based on a "pull" approach and user consumption. Toyota took this observation back to the manufacturing floor and, using a just-in-time manufacturing process, created the kanban approach to manage production based on the pull concept. This would keep inventory from piling up with no place to go. Excess inventory is waste and has a cost impact.

Although the kanban approach was initially applied in manufacturing, it is also applied in product development and software due largely to the efforts of David Anderson. Wherever kanban is applied, it is built upon a set of six defined practices (see Table 8.2).

Table 8.2: Kanban Practices Defined

Practice	Description
Visualize WIP	The team must be able to visualize the work and the process it goes through.
Limit WIP	The team agrees to limits around how much work can be in process at a given time. This helps decrease flow time.
Manage flow	Use observation and empirical controls to identify and address bottlenecks.
Make policies explicit	The team captures their agreed-upon policies such as WIP limits, definition of *done* at each stage, and other rules for working together. It is the agreement that determines when a work item in the kanban is ready to move from one column (state) to the next column (state).
Implement feedback loops	Obtain feedback from users and stakeholders on work completed
Improve collaboratively, evolve experimentally	The team continuously learns, innovates, and improves the state of the product and the process to improve flow.

A kanban board can be used to help manage the practices and provide the transparency needed to manage the flow of work. In the kanban example below (Figure 8.3), you can see where the WIP limits have been defined by the team. They can visualize the workflow across the different states of development, manage the flow of work (often as stories or tasks), and capture the policy for each column. An example card shows how a task might be captured by identifying what the work and tasks are, the importance and priority, and the estimated effort for the work.

During the daily planning horizon, as outlined in Chapter 5, team members read the board right to left, keeping aligned with the pull system. This team would observe a possible bottleneck starting to emerge in testing. Just as with inventory on the manufacturing floor, we want to avoid the pileup of work that sits with no place to go. In this scenario, the team would likely swarm to remove the bottleneck in testing. Teams iterate on designs and development while using a kanban board to visualize and manage the work.

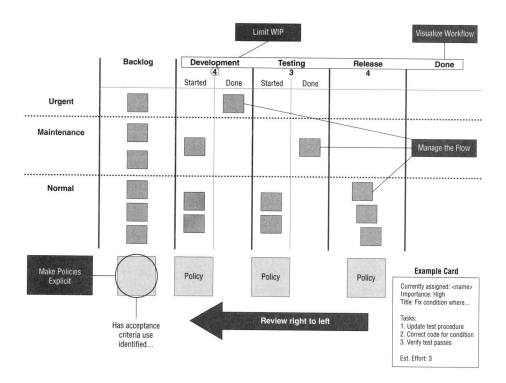

Figure 8.3: Example Kanban Board

One caution to watch out for when implementing kanban is a team that is using the board only as a series of states. There have been many occasions where we have witnessed teams moving from a time boxed approach such as Scrum to a single-piece flow approach such as kanban. However, in many cases, the teams are using the board only as a holding place with states for their work. Kanban is very disciplined and requires tracking of lead time, cycle time, WIP limits, and swim lanes that manage different classes of work.

Iterative Delivery for Large Systems (Teams of Teams)

Large, complex cyber-physical systems require the integration of capabilities across many teams of teams and suppliers. The number of people involved can be anywhere from a few hundred to thousands, depending on the type of system. To improve the flow of work, it is important that these organizations be organized around the systems within the value stream, not functions, to yield the benefits of iterative development and incremental delivery (see Chapter 4).

While iterative and incremental development is often thought to be for software development only, this was never the case. Many of these practices were originally discovered in product development. Bringing back the approach holistically at the system level to optimize development cycles is imperative. The maturity of digital tools and environments has further enabled these approaches. As the physical world moves into the digital environment, the cost of change is less, and the flexibility of change increases.

As described in the paper *Overcoming Barriers to Industrial DevOps,"* "during the exploration phase, teams build digital shadows (low-fidelity models) with simulation, emulation, and data flows. As the fidelity of the shadow increases, the entire product can be digitized and represented in a digital twin construct. As the product is built iteratively it is demonstrated and released in increments. An increment is functionality that is integrated and demonstrable."[10]

Continuing the example with our CubeSat mission, the teams are working to develop the attitude control functionality. There are multiple teams working together across functional areas to build, test, and implement the attitude controller. Each team is managing their workflow each iteration through the use of a kanban board (see Figure 8.4). The teams iterate to build out the algorithms to capture the spacecraft's position and orientation.

During this process, early iterations incorporate changes into the target software environment. The teams will demonstrate in a processor-in-the-loop environment that simulates the activity in software. With each iteration, they have built out an increased functionality that can be tested and demonstrated for feedback. The teams continue to iterate on the capability. In the next iteration, they incorporate the changes into hardware-in-the-loop closed-loop simulation.

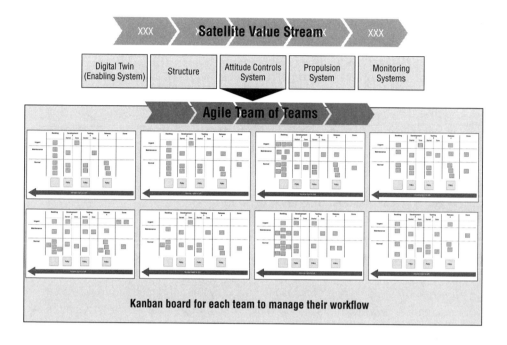

Figure 8.4: Visualizing the Flow of Value through Team Kanban

In early stages of product development, the hardware is not available. As a result, early iterations of development occur in virtual and simulated environments, and over time, development will include the physical environment as hardware becomes available.

The aerospace and defense industry continues to see results as it adopts and scales Industrial DevOps principles. For example, SpaceX successfully demonstrated their ability to quickly iterate and learn as they built the reusable rocket. Using models and 3D printing they can demonstrate successful results over and over again.[11] It has also been reported that SpaceX has realized cost savings by a factor of ten through investment in digital platforms and improved efficiencies.[12]

As was reported by *Harvard Business Review*:

CAD and finite element analysis software store rocket assemblies and databases that are shared across various teams of experts through a centralized repository. The software is fast, which encourages engineers to perform rapid iterations of both its virtual prototypes. The centralized nature of databases also promotes communication and collaboration of SpaceX's teams, thus removing silos.[13]

The US company Launcher has made use of additive manufacturing to accelerate their development of rockets with the mission to carry small satellites into orbit. It was reported that in April 2022 they successfully "demonstrated nominal thrust, pressure and oxidize/fuel mixture ratio for the first time with their 3D-printed E-2 rocket engines."[14] Developing this capability was something that took many iterations of testing and learning coupled with the use of their 3D printing capability to make this possible.

CASE STUDY: Alliant Techsystems Inc.[15]

At Alliant Techsystems (merged with Orbital Sciences and later acquired by Northrop Grumman), a team working on the launch-abort motor for the Orion space vehicle had an initial steel manifold weighing two thousand pounds. The launch-abort motor is part of the launch-abort system designed to pull the Orion crew module away from the launch vehicle in the event of an emergency on the launch pad or during ascent. The manifold uses a reverse-flow technology that forces hot gas through four nozzles at the top of the rocket, creating a pull force that would safely pull the crew module away from the launch vehicle should an emergency occur.

Due to a requirement to reduce the overall launch system weight, they decided to explore different options that could reduce the weight of the manifold while still being strong enough to handle the force required during launch. To address this need, the engineering team created software and simulation models so they could experiment and test a variety of solutions. Using simulation software for hardware and an iterative development approach, they drastically reduced overall design time while being able to innovate/design/test iteratively as they discovered an optimal solution meeting both weight and maximum load requirements.

The structural optimization software tool used was OptiStruct, which enabled engineers to digitally simulate hundreds of iterations and launch-abort scenarios using their supercomputer in Utah. The team was able to use this environment to optimize the geometry and design of the large titanium manifold and optimize the overall structure to carry the thrust, pressure, and line loads while placing material in the necessary and critical load paths. They blended the computer-optimized design with optimal machining practices to design and build the manifold. This approach saved hundreds of pounds in weight and titanium material while still being able to support the thousands of pounds in "pull force" from the four nozzles needed to lift the astronauts to safety should a problem arise during launch.

GETTING STARTED

- Define incremental small experiments to rapidly learn as you build your solution.
- Iterate on solutions to reduce uncertainty and reduce program risk.
- Proactively identify and manage queues that exist in your workflow. Use tools that can help your team "visualize" the bottlenecks and measure throughput.
- Set up a kanban board to manage and visualize the flow of value.
- Ensure you are trained on kanban as opposed to kanban theater.

KEY TAKEAWAYS

- Small batch sizes reduce queue lengths, enabling continuous flow and faster delivery.
- There is an economic opportunity cost experienced with long queues.
- Incremental experiments increase learning opportunities and innovation.
- For hardware solutions, simulation software and models enable faster learning cycles.
- Iterative implementation reduces uncertainty.
- The length of your iteration correlates to speed of learning.

QUESTIONS FOR YOUR TEAM TO ANSWER

- What experiments can you start with to increase your learning opportunities?
- Have you identified the queues that exist in your workflow?
- How are you visualizing and managing your WIP?
- As you build a hardware solution, what is your approach to learning and testing iteratively for fast feedback?
- At what frequency will you iterate? Why?

COACHING TIPS

- Michael Kennedy advocates that "everyone creates knowledge and acts on it for the good of the whole." Harness the learning and knowledge creation through short iterations and fast feedback loops.
- An incremental iterative approach allows the team to improve the state of the product and improve the state of the process with each iteration.
- Break down the backlog so there are small batch sizes of work that go through the system faster, reducing cycle time.
- Stop starting and start finishing.
- Engage your customers for regular feedback.

CHAPTER 9

ESTABLISH CADENCE AND SYNCHRONIZATION

> **PRINCIPLE 6:** Establishing cadence and synchronization discusses how these two concepts complement each other to manage variability and improve predictability.

It's time for the launch. Everything is ready for the countdown. The spacecraft, mission management team, test director, etc. are all ready for this moment. The next action is the main engine start. This is the ultimate synchronized moment, but it is not the only time that synchronization has happened in preparation for this event. Synchronization and events occurring on cadence are critical success practices in the cyber-physical world. When development and major events are not synchronized, the results can be late schedules and missed opportunities.

Cadence is like the rhythmic pattern of a heartbeat. Like a pulse, it is steady and happens regularly. It is like the sound of troops marching or the drumbeat in your favorite song or the steady, pounding rhythm of runners as their feet hit the concrete with each step they take. Cadence is what makes everyday activities and events predictable. Each day at 3:00 PM, the mail is delivered. Every school day morning at 8:00 AM, the bus stops at the corner, picks up the students, and carries them off to school. Every Sunday at 6:00 AM, fresh produce is delivered to the local market.

In the United States, Thanksgiving and the Christmas holiday season occur at the same time each year. This helps retail businesses in their planning and in the hiring of staff to accommodate the increased shopping demands. Companies plan for major releases of new products if they want to take advantage of the increased shopping and holiday market. Cadence helps businesses in the planning and release of new products and new product capabilities. Cadence makes routine that which can be routine (see Figure 9.1).

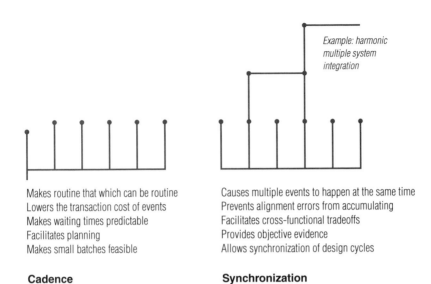

Example: harmonic multiple system integration

Makes routine that which can be routine	Causes multiple events to happen at the same time
Lowers the transaction cost of events	Prevents alignment errors from accumulating
Makes waiting times predictable	Facilitates cross-functional tradeoffs
Facilitates planning	Provides objective evidence
Makes small batches feasible	Allows synchronization of design cycles

Cadence   **Synchronization**

Figure 9.1: Cadence and Synchronization
Source: Josh Atwell et al., *Applied Industrial DevOps*.

In combination with cadence is synchronization. Synchronization is the integration of activities that are occurring on a cadence. Synchronization happens on cadence versus at ad hoc, random moments. The band members practice on cadence. They meet every weeknight at 7:00 PM for practice. When they practice their music, the goal is to synchronize their parts. Through the cadence of the music, each musician synchronizes their parts and the whole band delivers harmonious music.

For an organization's internal operations, this is like yearly strategic planning and goal setting with aligned objectives. Yearly objectives and road maps define budget allocations with quarterly planning and iterative planning cycles. Maybe you experience something like yearly operating reports and monthly budgeting reports at your organization. Within the iteration, your product teams plan their work, have daily stand-ups, and conduct demonstrations for customers and stakeholders, all on a regular cadence. Cadence and synchronization create predictable events for planning, feedback, and improving.

Organizations and teams can experience great benefits by incorporating these two practices. This is especially true when building cyber-physical systems, which require the coordination of many teams with different functions. Creating a shared cadence and synchronization helps people know when certain events will take place, lessens mental fatigue, and reduces confusion as clear expectations around the cycles are defined and understood. Recurring events create a "business rhythm," which becomes part of "how we do business." If product features are delivered on cadence, customers know and begin to anticipate the release of new functionality.

Planning (Cadence and Synchronization)

Let's reflect on the multiple horizons of planning (Industrial DevOps Principle 2). Each horizon of planning occurs on a regular cadence. Each defined cadence provides learning and insights to the levels below and above it. Each of these horizons is then synchronized in a cascading cadence (see Figure 9.2).

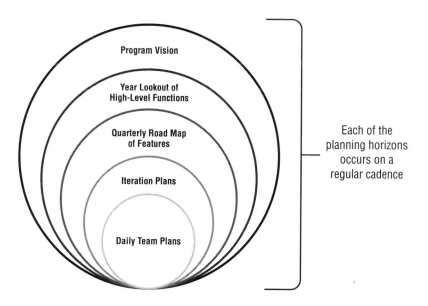

Figure 9.2: Cadence of Multiple Horizons of Planning

Product teams planning on cadence create synchronized plans resulting in the alignment around major product deliveries or milestone events (such as a flight test for an aerospace system wanting to test a set of functionalities for feedback). Synchronized planning ensures the highest-prioritized capabilities are aligned across the teams, so they know when they can integrate with other systems and that the work of the teams is aligned to deliver in the time needed. Therefore, the teams within each value stream have a defined and agreed-upon cadence to make this a reality.

The cadence depends on the needs of your organization. Consider how quickly your teams and organization want to go through the learning cycles and how frequently you want the opportunity to reprioritize. The quarterly cycle often aligns well with internal business rhythms, making it easier to adopt in parts of the organization new to their Industrial DevOps journey.

With large cyber-physical systems, there is an increased need for transparency and synchronization to ensure teams are building an integrated system of defined priorities. For example, with teams planning on cadence, product managers define the prioritized features and engage regularly with stakeholders for input. This is a coordinated time for them to ensure alignment toward common objectives and expectations as they are approaching the next quarterly planning cycle. As they define quarterly plans, teams collaborate to plan their work and determine the synchronization points and demonstrations.

Iteration planning also happens on cadence. Teams do not define their iteration cadence in isolation; their iteration cycle is synchronized with the cadence of other teams. This ensures defined integration and synchronization of functionality so the progress of the product can be demonstrated regularly.

When building cyber-physical systems, these practices extend across functions—that is, software, firmware, and hardware. Planning and aligning work on the foundation of cadence and synchronization enables teams to identify specific integration points for the development of functionality, whether it is done through a digital environment or models, and when they are looking for the early integration points, into the physical hardware and into manufacturing.

Example of Planning Cadence and Synchronization

Using the CubeSat example, the development teams are aligned around primary functions of the attitude control system (ACS). Their quarterly planning occurs on a regular cadence, the first week of the quarter.

The cross-functional teams discuss different iteration lengths. They know they need to plan on cadence, and they will want to be synchronized as they plan together to work toward developing shared capabilities. There are concerns to work through. The structure team has concerns about how to break down the work to a level that can fit within the iteration and still have a meaningful demonstration at the end. They understand a one-to-four-week iteration cadence is the norm in the industry, and that is their desire; however, the teams feel like it is too big of a stretch for them, as they are just getting started with Industrial DevOps practices. The team focused on hardware functions decides to start with a three-week cadence due to the fact the teams are new and need time to learn how to decompose their work into small enough pieces. They plan to assess again at the end of the quarter as to how that time box is working for them and if they are getting feedback fast enough.

Product owners refine and prioritize the team features on a regular cadence with their team. However, since the CubeSat ACS team has eight Agile teams, the product management team (team of product owners) meets on a regular cadence to refine and prioritize the features for the upcoming quarter. They decide to schedule this recurrence approximately three to four weeks prior to the next quarterly planning session. The following calendar (see Figure 9.3) represents the quarterly cadence defined by the CubeSat ACS team. Because events happen on cadence, they become routine and therefore can be scheduled.

Sunday	Monday	Tuesday	Wednesday	Thursday	Friday	Saturday
Quarterly planning with teams & customer						
			Iteration planning			
		Integrated demo	Iteration planning			

Sunday	Monday	Tuesday	Wednesday	Thursday	Friday	Saturday
			Integrated demo	Iteration planning		

Sunday	Monday	Tuesday	Wednesday	Thursday	Friday	Saturday
Feature refinement for next quarter	Integrated demo	Iteration planning				
		Integrated demo				
	Quarterly Planning with teams & customer					
				Iteration planning		

Figure 9.3: Example Schedule Based on the CubeSat Mission

Product Development Synchronization

Cadence must be combined with synchronization. Synchronization on cadence provides the organization with not only the opportunity to know when planning events and demonstrations will happen but also insight into how the whole system is progressing (see Figure 9.4). Synchronization provides feedback at every iteration by answering these questions: Did we build it right (validation)? Did we build the right thing (verification)? Together, cadence and synchronization unite to create the opportunity for continuous learning, demonstration of the product, and a feedback cycle.

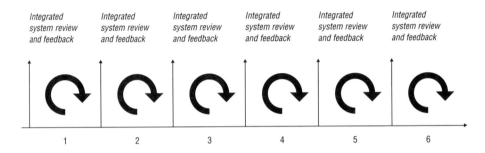

Figure 9.4: Synchronized Cadence

Having synchronization at each iteration reduces variability between what the system is expected to do and how the system is actually performing. At the systems level, each iteration results in an integrated systems-level review and the opportunity to adjust if the new functionality is not performing as expected. This reduces the overall variation from what is desired to what is delivered. Iteration cycles that are too long may result in greater variance and negatively affect cost and schedule.

Together, cadence and synchronization define standard practices, enabling the rhythmic pattern of repeatable planning sessions, integration, and demonstrations of integrated working capabilities. With large cyber-physical systems, these demonstrations may initially be performed in the model and then later integrated into the physical hardware.

For example, synchronization across a quarter means:

- Iteration planning and demonstrations at the team level are defined on a repeatable cadence.
- Quarterly planning is defined on a repeatable cadence.
- Integrated system review/demo and feedback occur at each iteration. This happens on a defined cadence and is the synchronization of the system. Having regular synchronization reduces risks and provides a regular feedback loop on how the system is progressing.

Let's visualize what happens when cadence and synchronization come together.

Cadence and Synchronization Combine

Teams are planning and developing together to build the attitude controller and determine attitude position. Each iteration is a synchronization point for the teams to better understand how the system is performing. This means the teams are aligned around the system they are building and are all marching to the same iteration cadence. The various teams work together to determine what testing is done in simulated or emulated environments. While the process of how to do the work is happening on cadence, the act of building and demonstrating their work is synchronized.

They plan on a regular cadence (Principle 2) and synchronize on cadence through short iterations (Principle 5) to demonstrate working functionality (Principle 3). Integrating these Industrial DevOps principles, the CubeSat teams can improve the flow and delivery of value through cross-functional planning and execution.

Large cyber-physical systems typically have multiple teams designing, implementing, and deploying multiple interconnected subsystems and components over long periods of time. The teams will uncover unknowns, and they will strive to exploit good variability and remove bad variability. The trick is knowing which is good variability and which is bad. Cadence and synchronization are two tools that aid in quickly uncovering variability in the system.

The Challenges

Let's visualize what happens when cadence is not synchronized. What does that look like? Imagine that you are working with multiple teams building a cyber-physical system. Potentially, the structures team for a vehicle were not using Agile time boxing before, and insist they need six weeks before they can demonstrate progress on the physical structure. On the other hand, the GNC team has been using time boxing for much longer, and two weeks has been working well for them. Together the teams collaborate to synchronize on an integration point schedule.

The teams are executing on cadence, just not the *same* cadence. This makes regular synchronization and demonstrations difficult. At best, the teams will be able to demonstrate integrated capabilities every six weeks, when it is time for the structure team's demonstration.

While this might be good enough in the short term as teams are learning how to work in this new environment, this scenario presents a missed opportunity for more frequent feedback loops to learn and improve, resulting in increased delivery risks. Synchronization is absolute at launch time, but its value needs to be extended on a frequent cadence to reap its benefits. While teams new to this approach may consider this to be a "good enough" starting point, they will want to regularly review their processes and identify how they can improve.

Conclusion

In this chapter, you were introduced to the importance of cadence and synchronization. These two concepts are closely intertwined and must be aligned across teams building an integrated capability to yield the desired outcomes. It increases the frequency of feedback loops and increases the opportunities for regular improvements, leading to higher quality and building better systems.

GETTING STARTED

- Create synchronized plans resulting in the alignment around major product deliveries or milestone events.
- Create a common cadence for your team to plan and demonstrate an iterative learning and feedback loop from stakeholders and users of your product.
- Synchronize cadence across teams when building an integrated solution that is demonstrated to stakeholders and users.
- Validate that frequency of your feedback loops enables learning.

KEY TAKEAWAYS

- Cadence is the rhythm; it is the pulse of the business organization.
- Synchronization is the touch point where things come together to verify and validate the system. When applied to short iterations, it reduces the variance of the schedule, and product teams can quickly pivot when different design considerations are needed.
- Each iteration is synchronized on a defined cadence; therefore, when doing quarterly planning, a team can identify the dependencies, the

interfaces, the use cases to be worked on, and the integration that can be demonstrated.

QUESTIONS FOR YOUR TEAM TO ANSWER

- What planning horizons did you select for your product? Why?
- What cadence did you select for your teams?
- Are your teams synchronized on the same cadence?
- How often do you need feedback? Have you discussed the cadence with the stakeholders you expect to participate?
- What is considered good variability in your context?
- What is considered bad variability in your context?

COACHING TIPS

- Define your program or product team's cadence and synchronization points. Cadence is the rhythm and makes recurring events predictable. Identify any existing cadences in the organization that you can build from.
- Synchronization refers to the synchronization of plans, development, and demonstrations. Have well-defined interfaces and synchronization points with your suppliers.
- Once a cadence is defined, schedule the recurring planning and demonstration events for the next six months to a year. You will find that teams will plan around these recurring events so they can be present and engaged. Change the cadence only when it is necessary and communicate to the teams why the cadence is changing. Stay on cadence to reduce stress on the team.
- To realize the benefits of synchronization, your teams must be organized around the value stream. Review your documented value stream. Do you need to make any adjustments?

CHAPTER 10

INTEGRATE EARLY AND OFTEN

> **PRINCIPLE 7:** Integrating early and often covers different levels and types of integration points across large, complex systems to improve speed of delivery.

System integration is the process of combining all of the components and subsystems into a unified system. The process requires linking software, firmware, and hardware to create a seamless flow across the system. The goal of integration is to enable demonstrations for decision-making, improve flow, work in smaller batches, and ensure the systems work together effectively.

Until we have integrated the system, we have a lot of risk and uncertainty of whether the system will be fit for purpose. With many decades of system development and delivery between us, we both have multiple stories where the schedule and cost for the system were looking really positive, but the day after our first integration, we realized just how far behind in both cost and schedule we were. There is a direct correlation between the length of time between integrating and the amount of rework that is required.

For commercial companies delivering products, continuous integration (CI) is the gold standard. CI is the practice of automating the integration of software capabilities from multiple contributors into a single integrated baseline for the end-to-end system. This practice requires extensive discipline and investment into automation. Companies such as Netflix have invested millions of dollars into CI platforms that allow them to safely integrate software products in subsecond times.

As we scale, we have to consider constraints associated with physical products. For cyber-physical systems, our goal is to integrate as frequently as possible to evolve the system to meet customer needs and get feedback on what is working and where improvements can be made. Some considerations that impact integration are investment of automation, lead times in hardware, expensive test equipment, regulatory compliance rules, and training of employees on the tools and processes.

CI beyond software development may be a step too far for some companies building cyber-physical systems. However, recent advancements and adoption in technologies such as modeling, simulation, digital twins, and additive manufacturing are enabling organizations to move physical systems into the digital space, which opens opportunities to get closer to CI previously leveraged for software development and understand how it will perform when it is integrated with the physical components. A recent workshop we attended shared that SpaceX can flow software changes to a high-fidelity integrated hardware lab environment in under twenty-four hours.

Digital integration cannot take the place of physical integration, but it does reduce risk exposure while providing the ability to evolve and learn faster. Using models and digital twins allows teams to integrate in the digital space, which provides visibility into system development. Another maturing technology that allows faster integration is additive manufacturing, which gives us the ability to 3D print components of the system.

Relativity Space was founded in 2015 with the goal of building rockets faster. Their flagship rocket, Terran 1, has been designed to be entirely 3D printed. This approach allows them to rapidly evolve, iterate, and integrate in cyberspace with a digital representation of the vehicle and print the finished product.[1]

Integration with Software

CI for software is the process of integrating software into a single end-to-end baseline multiple times a day. CI identifies and resolves issues rapidly. With each check-in, a multi-tiered suite of automated tests verify and validate the update to ensure it meets the intent. It achieves this by creating a flow of delivery and receiving fast feedback on how the system is working. By building a CI pipeline, we achieve a multidomain flow of value that can be deployed (see Figure 10.1).

At a thousand-foot level, the process begins when the team begins to code, places the software under version control, commits to a CI environment, and goes through a series of testing and validation assessments. Then the code is approved and released into production. One barrier pattern seen is teams creating long-lived branches while claiming CI. If you are seeing branches lasting longer than a day, then by definition, they are not implementing CI.

This framework requires three capabilities to work:[2]

1. A comprehensive and reliable set of automated tests that validate we are in a deployable state
2. A culture that "stops the entire production line" when our validation tests fail
3. Developers working in small batches on trunk rather than long-lived feature branches

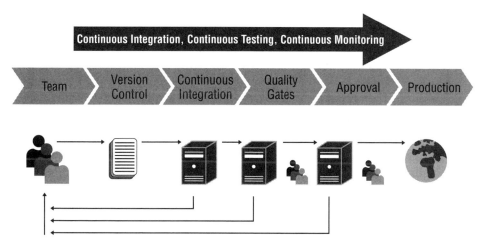

Figure 10.1: CI Pipeline to Optimize the Flow of High-Quality Features to Users

In the book *Continuous Delivery: Reliable Software Releases through Build, Test, and Deployment Automation*, Jez Humble and David Farley lay out some basic practices of CI for software. The team has coded their software, and they are ready to check in their latest updates. The basic steps are outlined below:[3]

1. Check to see if the build is running. If so, wait until the tests have finished, and when the system is ready, check your code into the development environment.
2. Execute the build script and run your tests with the configuration tool.
3. If the build fails, fix it. When it passes, move to your next task.

Digital natives, such as Amazon and Google, have been successfully implementing CI and automated testing in software environments for decades.

Integration with Embedded Systems

Several studies are being performed with CI in firmware for embedded systems. One study, released in 2019, defines a continuous integration/ continuous delivery (CI/CD) process for firmware development.[4] Their model emphasizes the importance of CI across five stages and different forms of testing that include unit testing (UT), build verification testing (BVT), basic acceptance testing (BAT), and nightly test environment (NTE). These tests are an integral part of their CI/CD pipeline, going from branch commits and daily builds through regression testing to a release process. Through a CI/CD approach and improved test automation, the study showed improved quality and efficiencies in identifying and fixing issues related to development and hardware changes.[5]

Figure 10.2 provides an illustration of the five stages of a CI/CD pipeline for embedded systems. While the illustration highlights first article inspection (FAI), we recommend an alternative approach. Instead of FAI in the traditional sense, quality is built into the iterative process, making inspection an ongoing activity within each iteration versus bolting on inspections at the end. This will result in higher-quality products with less cost because you have reduced the amount of rework and the hand-offs.

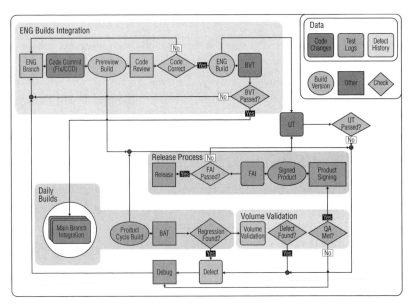

Figure 10.2: CI/CD for Firmware Development for Embedded Systems

Source: Mateusz Kowzan and Patrycja Pietrzak, "Continuous Integration in Validation of Modern, Complex, Embedded Systems."

Integration with Hardware

Iterating on hardware designs and leveraging the repetitive development inspired by set-based design enable teams to explore multiple considerations and systematically reach the desired target (see Chapter 8). This emphasizes learning and exploration. The iterative approach enables teams to experiment, learn, and converge toward an optimal solution. Part of the iterative and experimentation process means integrating and testing solutions as frequently as possible. Some of the early learning may be through simulated environments. It may involve prototypes or testing new capabilities on earlier versions of hardware to improve the software/hardware interface. Through each iteration of software and hardware development, the goal is to integrate as early as possible, discovering what works and what doesn't work and keeping a capability that can be demonstrated.

With software capabilities, teams are implementing a CI approach with the expectation of integrating daily with the embedded software. With the introduction of teams who are building hardware components, teams need to evaluate the technologies and digital capabilities available, along with the financial investment needed to build out the digital ecosystem and create an approach that fits best in their environment (see Figure 10.3).

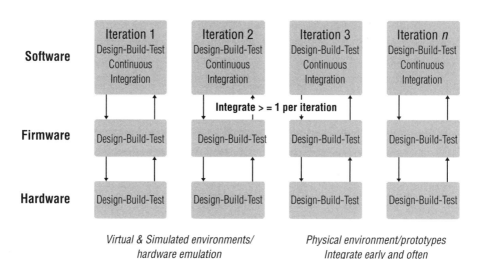

Figure 10.3: Software/Hardware Integration

In some cases, especially earlier in development, the new hardware may not exist yet. However, even if the new hardware doesn't exist, cyber-physical teams still have options. Teams may test their new designs and capabilities with prototypes, existing hardware, models, digital twins, digital shadows, or other simulated environments. Digital twins are evolving rapidly with the design and building of cyber-physical solutions. The digital twin is a companion to the physical solution yet exists only in the virtual world, spanning the entire life cycle of the system. It is a "set of models that allow the digitized physical object to be viewed in three dimensions on the computer. The models are very critical in the DT [digital twin], since they developed the initial concept 'twin' into 'digital twin' and provided more deep insights through simulation."[6]

Integrating with Manufacturing

Development teams engage with manufacturing the same way they engage with operations. Cyber-physical teams must first remove functional silos, which includes embracing the relationship with manufacturing. Teams design and develop iteratively, then integrate and test in hardware. Through these small experiments and tests, development teams engage with manufacturing by applying small batches and a feedback loop from design to manufacturing during the development of the new hardware or enhancements to the physical hardware.

For example, manufacturing is a key stakeholder for product development teams when procedures and design specs are written. Manufacturing representatives engage with the Agile teams to identify early integration points and to communicate risks and opportunities. In some organizations, at this phase of design and development, manufacturing may have team members embedded within the Agile development teams. The integration of development with members of manufacturing provides the opportunity for early integration and testing of products, including working instructions used by manufacturing, and reduces rework from occurring further downstream.

The digital twin capability helps with exploratory integration and testing of new capabilities and the analysis of how the physical solution responds or performs in different situations. With the evolution of smart manufacturing and digital twin capabilities, the engagement between development and manufacturing continues to mature and slowly become more integrated.

Example

Let's look at a notional example with our CubeSat team and some of their integration activities. In this scenario, they are working on two capabilities. One capability is a software-only update to the attitude control system. The second capability includes a new image sensor for the weather satellite for improved resolution. With both scenarios, a few possible integration activities for the capabilities have been included. Please note that this is not exhaustive of the integration activities that need to occur in the development of a real system.

Capability 1: Software-Only Updates to the Attitude Control System

- The team develops the changes. As they develop the new software, they are implementing the DevOps foundational practices of CI and automated testing.
- Teams can deploy software updates to hardware if available or leverage their digital twin and test environments; code-level integrations can happen routinely, daily, or even hourly. There was an initial up-front investment in these environments; however, the return on investment has been worth it. Hundreds of automated tests are run daily, ensuring built-in quality.
- DevOps practices of source code control, automated builds, and automated build verification tests are applied.
- The team has integrated tools to see when a build breaks and needs attention.
- After testing and validation of new software enhancements, updated software is released to the existing satellite.

Capability 2: New Image Sensor for a Weather Satellite

- The team is working on a new camera for the next satellite, and the specs are defined.
- The image sensor has been simulated, which feeds simulated data in via the predetermined API and protocols. The team deploys the device into the test bed. In the test environment, a new algorithm is developed for precise spatial resolution, Earth imaging, and mapping without the use of a physical device yet. They also employ a testing mock prototype.

- The mechanical design is tested with mechanical mock-ups, which are consistent with the intended physical properties of the image sensor. The sensor hardware is integrated in the test environment, and the new capabilities are tested on hardware.

The Industrial DevOps principle of right-sized integration needs several other Industrial DevOps principles for better results:

- Principle 1: Organize for the flow of value around value—teams must be organized to optimize flow.
- Principle 3: Implement data-driven decisions—we always want to see something working.
- Principle 4: Architect for change and speed —this requires a loosely coupled architecture to enable the ease of integration.
- Principle 9: Shift left—this requires commitment to automated testing and a test-first mindset.

Integration Strategy Considerations

Developing an intentional robust integration strategy for your cyber-physical system needs to be a high priority to enable, if not continuous, then at least frequent, integration. This is often an afterthought once we begin building the system, which impacts the level and frequency of integration possible. Begin with those high-level use cases and test cases to determine the tools that are needed to ensure this happens. For example, say you want to test that you can command the satellite hardware to adjust the attitude using a software command. You will need a lab environment that can simulate an actuator and the simulator to validate they integrate successfully, as shown in Figure 10.4.

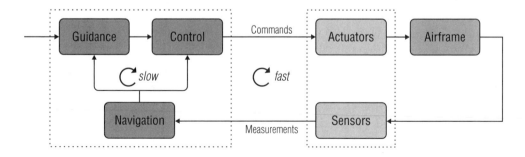

Figure 10.4: Cross-System Integration for Satellite

It's important to focus on the highest-risk interfaces, which typically interact with multiple subsystems. With our CubeSat example, we would heavily focus on the power, communication, attitude control, payload, and thermal interfaces. It would be important for us to set up the lab to frequently integrate and test those interfaces to drive down risk.

While there is no one-size-fits-all solution, the more frequently you can integrate changes across the system, the lower the risk. For example, as mentioned earlier, a recent presentation that we attended shared that SpaceX can integrate software changes in their hardware simulators in less than twenty-four hours. To determine the best frequency for your system, you need to perform a cost-benefit analysis.

We recommend building out your CI/CD pipeline before building anything. For example, for software applications we would download an open-source pipeline from GitHub and run something simple, like "Hello World." From there, each day we would add new capability to the pipeline. For cyber-physical systems, we would instantiate our simulation and emulation environments and begin running scenarios on simple models.

There are multiple types of lab environments to consider, depending on the type of product you're building, such as hardware, environmental, mechanical, power, prototype, and high-fidelity integration. Hardware labs support testing and validating equipment such as oscilloscopes, multimeters, signal generators, and logic analyzers. Environmental labs enable the validation of temperature, humidity, vibration, and shock resistance. Mechanical labs address durability, stress resistance, and load capacity. As the name suggests, power labs validate power consumption, voltage regulation, and thermal management. Prototype labs allow you to test early-stage prototypes like those that are 3D printed. The high-fidelity integration labs are the most expensive but often the most critical, allowing us to validate the interaction of hardware components such as sensors, actuators, and control systems.

CASE STUDY: Intel Technology, Poland[7]

Intel, like other companies, has experienced increasing levels of product complexity, where advanced systems include hardware, firmware, and software. Intel has numerous siloed teams who individually provide those components. To further exacerbate the problem, each of these teams is in a different location, and their schedule of releases for the components is not published. In addition, these teams have very expensive validation processes, so there is a belief that there is room for cost optimization. The critical challenges Intel is attempting to address are scale and complexity, high-volume code change, multiplicity of environments, and limited resources in areas of devices under testing, workload and traffic generators, measurement setups, and test probes.

Intel experimented by instantiating a CI/CD pipeline that included five stages of testing, which were unit, build verification, build acceptance, nightly, and first-article inspection tests. The team put a great deal of effort into building automatic self-recovery procedures in case of failure and ensuring proper redundancy. During the process, they gathered a full suite of metrics to determine levels of success.

The results of CI and CD provided an array of benefits for the team, including faster and more efficient defect detection, improved quality, and improved development and validation performance. Due to the success of the experiment, the team has identified further improvements they can make to the pipeline to increase the level of results.

CASE STUDY: Vehicle Research Center: A Metadata Model for DevOps[8]

The virtual vehicle research center in Graz, Austria, performed research on developing a CI pipeline for automobiles due to the increasing complexity of the systems and the need to stay ahead of quality. Automobiles have become computers on wheels, and both the number and complexity of the software functions have dramatically risen. In 2010, a car contained roughly 10 million lines of code, today's non-automated cars have over 100 million lines of code, and tomorrow's automated cars are expected to have between 300 and 500 million lines of code.[9] As a result of the growth, old ways of testing are no longer feasible and only get worse with autonomous driving capabilities.

One of the most popular ways to make large systems development and testing go more smoothly is through CI. CI is a software development practice where teams integrate their work very frequently, and each integration is verified by an automated build with a full suite of tests. This approach has been used for pure software development, but new techniques need to be identified to work with a co-simulation environment. Complex simulations require extensive resources in terms of hardware and time; in addition, they require techniques to enable the composition of simulators to support. A general simulation tool cannot be used due to the number of varied simulation domains involved. To support this, the team leveraged a graph model with graph-based metadata. The team built a simulation development chain where components were deployed with the help of autogenerated build pipelines from graph-based metadata.

The result was that the CI labor efforts were reduced by providing tools and a framework for the automated generation of CI pipelines. Instead of creating all CI scripts for the co-simulation by hand, a metadata model, which was stored within a graph database, was used to autogenerate all the pipeline data needed for a co-simulation. The results were positive and demonstrated that this approach could be applied to the early stages of development. The team plans to extend this research to more complex use cases, optimization improvements to test automation, and investigation of SysML to enhance the metadata model.

GETTING STARTED

- Define high-level integration vignettes and scenarios.
- Define high-risk areas.
- Build and invest in a CI and CD pipeline for software.
- Determine high-level integration frequency.
- Develop multi-tiered automated tests (unit, integration, regression, performance, and security).
- Develop a high-fidelity simulation pipeline to mock out hardware.
- Develop prototype labs to design, build, and test early-stage hardware prototypes.
- Invest in high-fidelity labs early with printed circuit boards and microcontrollers.
- Automate everything that you need to do more than three times.

KEY TAKEAWAYS

- Rightsize integration for software and hardware to reduce schedule and performance risk and improve flow.
- CI is a must for software.
- Define the integration approach for your cyber-physical systems. Consider software and hardware integration points.
- Identify what you need to improve the frequency of integration. Identify how the needs might be addressed.

QUESTIONS FOR YOUR TEAM TO ANSWER

- Have you defined high-level scenarios for your system?
- Have you identified key interfaces?

- What kinds of risks exist for your system?
- What items are impacting integration frequency?
- What is your strategy for standing up a CI/CD pipeline?
- How will you engage with manufacturing? What feedback needs to occur between development and manufacturing?
- Which labs do you need for your system? Why?
- What level of investment will be needed to build out your digital infrastructure?

COACHING TIPS

- Once you understand your value stream, define your delivery pipeline. You will need to capture the CI/CD for software and then the integration points with hardware development.
- CI is well understood for software. However, with cyber-physical systems, this becomes more complicated as we bring in hardware development and testing and manufacturing. With each quarterly planning horizon, plan how software and hardware integration and demonstration will happen. At first, this might all occur in models and then shift as hardware becomes available. Sometimes you might use prototypes or similar hardware for early testing of capabilities.
- Decide how and to what extent manufacturing is part of your integration strategy.
- Are you interested in learning more about the foundations of continuous integration and continuous delivery? We recommend reading *Continuous Delivery* (mentioned earlier in this chapter) and *The DevOps Handbook*.

CHAPTER 11

SHIFT LEFT

PRINCIPLE 8: Shifting left emphasizes a "test-first" mindset encompassing the multiple levels of testing across cyber-physical systems.

The Ariane 5 rocket experienced catastrophic failure shortly after the launch of its maiden flight in 1996. The failure was attributed to a software error in the rocket's inertial reference system, which measures the rocket's orientation during flight. The failure was a result of inadequate testing.[1]

That was nearly thirty years ago and just at the beginning of the dot-com era. Since then, software-intensive solutions have grown exponentially. As a result of this growth, the traditional ways of testing have become insufficient. Now, with the emergence of digital transformation and AI, there are even greater advancements coming that we have yet to imagine. For example, in 2018, Gartner predicted that by 2022, 40% of application development projects would have AI-powered developers on their team.[2] They posited that human developers would need to learn the role of translator by turning conversations with stakeholders and end users into precise tests and logic for their AI-powered co-developers. While we have not yet seen that rate of change in development team composition, ChatGPT has given us the first glimpse of the wave of the future.

Technology advancements are adding to mounting change and growing complexities. This has emphasized the importance of end-to-end testing and a "shift-left" mindset for software solutions. (In other words, testing and security are built in from the beginning.) However, the shift-left mindset that has been learned through the rapid development of software solutions can also be applied to the development of cyber-physical systems.

At this point, you may be wondering what is meant by a shift-left mindset. In systems engineering, "shift left" refers to the practice of shifting processes from the right-hand side of the Vee-model development process to the left (Figure 11.1), including areas such as testing and security. In other words, instead of waiting until large portions of the solution have been built to add security features or to think about how features will be tested, systems are built with the end in mind. Before development begins, teams consider how they will address needs such as security features and testing and what environments features will be tested in.

Begin with the End in Mind

Beginning with the end in mind is like having the answers to the test questions before you take the test. This mindset enables teams to think through how the capability will be tested before development begins. Typically, teams use this as part of the iterative approach as they develop in small batch sizes. Traditional development practices for cyber-physical systems meant having large batches of capabilities before they were integrated and tested, which can result in extensive and costly rework as defects are found much later in the development cycle. In order to deliver value sooner, it is essential to minimize rework and build quality and security into the development life cycle. This is even more important in cyber-physical systems, where rework means high cost and the loss of years of development work.

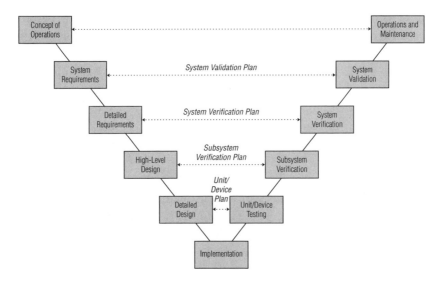

Figure 11.1: Vee Model

The Vee model is a development and testing model commonly used in the development of complex systems, such as cyber-physical systems. It is often associated with a waterfall life cycle, a sequential and linear process, as we illustrated earlier in this book. The model is named after the shape and defines a sequence of activities, beginning with requirements and ending with tests. The goal of the Vee model is to produce rigorously developed solutions; however, because of its pictorial representation, where testing is viewed last, many development efforts leave testing until the end, inviting costly and time-consuming rework.

Developing cyber-physical systems using the Vee model equates to designing and building the system first and testing after. The representation of the Vee model as shown violates best practices such as small batch sizes and minimal hand-offs. It also lends itself to a role-based or function-oriented organization. When organizations use the Vee model, large efforts tend to have one group gathering requirements, another group interpreting the requirements and creating a design, another group implementing the solution, and yet another group testing the implementation against the requirements (see Figure 11.2). While we recognize it does not have to be implemented in that manner, that is often the interpretation.

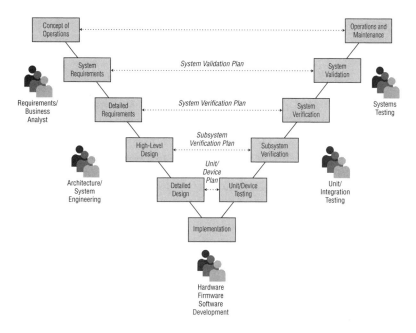

Figure 11.2: Vee Model Teams

Agile Requirements

This interpretation of the Vee model has often resulted in multiple teams interpreting language-based requirements, but written language is often insufficient in clearly articulating the vast and complicated requirements of a cyber-physical system. Written requirements are interpreted through each individual's or team's mental model. We all have very different sets of experiences that impact our mental models, leaving a wide variety of interpretations, as illustrated in Figure 11.3. The written requirement seems simple enough: "The system shall be a gray box." But the interpretation by each team shows the variety with which even the simplest requirement can be interpreted.

In this situation, teams using a traditional requirements approach would not learn that the development team had interpreted the requirement incorrectly until the end of the very long and costly development phase. In the case of something like a satellite launch system, this could result in missing a launch window and delaying the project by months.

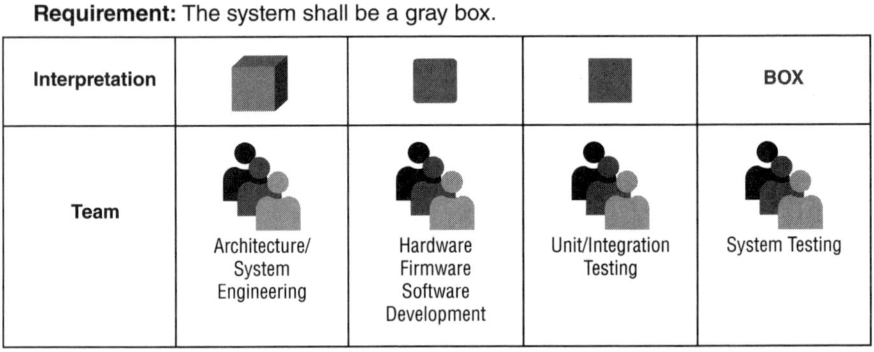

Requirement: The system shall be a gray box.

Interpretation				BOX
Team	Architecture/ System Engineering	Hardware Firmware Software Development	Unit/Integration Testing	System Testing

Figure 11.3: Requirements Interpretations

This also means rethinking how we define the backlog of work performed by the teams. For instance, if we bring this back to our CubeSat example, instead of having a *shall* statement that says "The attitude control system shall be able to transition smoothly and correctly between different control modes, including nominal, detumbling, safe, recovery, and emergency modes, in response to different scenarios and fault conditions," a team might have a set of user stories that would define this functionality.

For example, "As a CubeSat operator, I want the attitude control system to maintain a stable and accurate orientation of the spacecraft so that I can achieve my mission objectives." The story would include a test: "*Given* that our attitude control system is in nominal mode, *when* the spacecraft becomes unstable, *then* the attitude control system switches to detumble mode."

This format is important, as it causes the development team, which includes those who test, to think about the testing scenario before development begins. The *given-when-then* approach is one demonstration of how to write acceptance criteria, which are defined by the team before the work begins. At the features level, the acceptance criteria and test approach are defined before user stories are worked on by the team. This level of definition for the feature provides clarity around the scope of the work, how it will be demonstrated against acceptance criteria, and any other specific testing requirements that might need to be met. There are many testing practices required by teams building cyber-physical systems. The organization and teams identify their testing strategy, which includes shifting left as many of the testing practices as possible to continuously build in quality. In the sections that follow, we have captured many testing practices that organizations and teams should consider.

Automate Everything

In the software community, the mantra for decades has been "test early, test often, automate." As we move into the cyber-physical world, the amount of testing that is automated for cyber-physical systems can vary, but in general, automation is becoming critical for effectively testing cyber-physical systems to improve efficiency, accuracy, and consistency in testing. According to an article in Coders Kitchen, SpaceX has only fifty developers building software for their nine vehicles.[3] This is far fewer than traditional space programs, which use hundreds of developers. Automating is critical to being able to not only build but also test their systems. In addition, they use commodity components (like x86, unhardened PPC processors, and Linux), which enables a single workstation to simulate every controller and processor and allows them to scale automation. SpaceX claims they can push software into products 17,000 times a day.[4] Building a fully automated development pipeline with a full suite of tests ensures a safe and seamless integration. Based on our learning, we recommend automating any process you have done more than three times.

Challenges for Testing of Cyber-Physical Systems

Cyber-physical systems bring a new set of challenges to testing, which begin with increased levels of complexity where we have a variety of software and hardware components to bring together, as well as a range of variables to consider, depending on target environmental conditions. These systems often require expensive environments to test, including physical space, reliable power, and specialized equipment. Real-time performance is a big consideration for cyber-physical systems, where we often have very strict timing constraints with limited resources in memory, processing power, and battery life. Many cyber-physical systems such as automobiles, space systems, and medical devices, have extensive regulatory requirements with stringent safety and security impacts that must be integrated and tested throughout the product life cycle. Because of the safety and security needs, testing cyber-physical systems requires collecting and storing large amounts of data to validate against a wide range of scenarios.

System Complexity

The complexity of cyber-physical systems arises from the integration of physical and computational components. Today, typical cyber-physical systems are increasingly data driven, software defined, and hardware reliant. While efforts are made to reduce dependencies through the development cycle of hardware and software, eventually all components must come together to be fully integrated and tested, which is not trivial at scale. The impact on tests is extensive due to this heterogeneous nature, which involves integration from multiple vendors with varied technologies, creating challenges in interoperability and compatibility.

Think about our frustration with simple things like different power adapters. Consider the experience of someone who travels to a different country and needs different power adapters for the multiple devices they carry. Or when you are in a meeting and your mobile phone needs charging, you forgot your charger, and no one in the meeting has the right adapter that will work with your phone. This challenge of interoperability is magnified in cyber-physical systems. In some cases, these systems are distributed like power grids, making it very difficult to test the system as a whole. Well-defined interfaces and simulated environments can help address the complexity as the end-to-end system is tested. The larger the system, the more uncertainty that exists. It is difficult to test for emergent behavior that can arise in dynamic environments.

For example, once launched, satellites are in a highly volatile environment, and as reported by the European Space Agency, anomalies have been discovered with five out of eighteen satellites in orbit. While they have not identified the specific cause of these anomalies, they believe it may be short circuits and possibly the test procedure itself performed on the ground. It is challenging to address all the varied scenarios some systems will encounter, and the level of testing is correlated with the level of risk to safety and security.[5]

Lastly, the interdisciplinary nature of the physical and computational components creates another level of complexity that requires collaboration and coordination across different types of teams with different expertise. Do not underestimate the level of this problem. The teams needed to build the system almost never have a common mental model to communicate, leading them to talk past one another. As we shift testing left and begin with the end in mind, the complexities of the system must be considered and accounted for in your testing strategy and how you can work through these challenges through iterative development and testing cycles.

Testing Environments

To implement a shift-left testing mindset and address the complexities of cyber-physical systems, you must invest in several environments at the start of design and continue to build the environments throughout the cycle. Traditionally, lab environments have been built later in the development cycle, since in the waterfall approach, testing happens well after the design phase. This resulted in bottlenecks in the test labs. Specifically, new initiatives need to plan and budget for both software-in-the-loop and hardware-in-the-loop.

Software-in-the-Loop

A software-in-the-loop (SIL) is a type of testing environment that is used in early-stage tests. In an SIL environment, software components are tested in either a simulated or virtual environment, the hardware components replaced by models or simulations. The purpose of an SIL environment is to test the behavior of software components in a realistic context, allowing teams to test in a controlled environment before moving to the actual hardware. This type of testing allows us to simulate the inputs and outputs of the system in a low-cost environment. Integrating an SIL into the automated test pipeline requires setting up virtual sensors and actuators. The simulated environment can be adjusted to test a wide range of scenarios, including normal operation, edge cases, and failure modes.

If we use our CubeSat example, our teams would simulate the spacecraft using a combination of MATLAB and Simulink, including sensors and actuators running on a workstation, shown in Figure 11.4. We could begin the simulation with our real-time simulator (Simulink), validate our attitude control system was in nominal mode, and introduce a failure through the workstation to the attitude sensor, then validate the attitude control system transitions to detumbling mode.

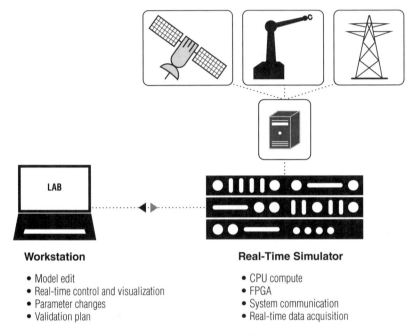

Workstation

- Model edit
- Real-time control and visualization
- Parameter changes
- Validation plan

Real-Time Simulator

- CPU compute
- FPGA
- System communication
- Real-time data acquisition

Figure 11.4: Software-in-the-Loop

Hardware-in-the-Loop

Hardware test beds are time-consuming and require hundreds of thousands of dollars to build. Hardware-in-the-loop (HIL) involves testing the hardware and software together. In an HIL lab, the hardware components of the system are connected to a simulation environment that emulates the behavior of the system's software components. This HIL environment, illustrated in Figure 11.5, can be used to test the hardware components under a range of different operating conditions and scenarios, enabling the testing of the system holistically.

To further expand on our test of the CubeSat, once we move from the SIL test to the HIL test, we connect the hardware-component attitude sensors, reaction wheels, and thrusters to the simulator. We could begin the simulation with our real-time simulator (Simulink), validate our attitude control system was in nominal mode, and introduce a failure through hardware, then validate the attitude control system transitions to detumbling mode.

The NASA Double Asteroid Redirection Test (DART) mission, led by Johns Hopkins, had big challenges in schedule and cost, which led them to follow a different path than typical space development. DART brought together many of the components of Industrial DevOps, including instantiating a DevOps architecture that integrated with NASA's Evolutionary Xenon Thruster—Commercial (NEXT-C) electric propulsion system. DART's DevOps architecture consisted of an SIL environment that simulated the spacecraft on a single laptop. Plus, HIL test beds were used with the goal of achieving a nightly simulated asteroid impact. The OV-1, a high-level concept graphic of the Johns Hopkins Test Environment, is illustrated in Figure 11.6. They leveraged a range of test beds, spanning from medium to high fidelity, to accomplish their mission.[6]

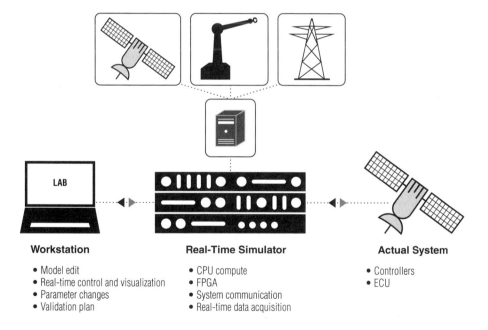

Figure 11.5: Hardware-in-the-Loop

The DART mission was able to successfully change the trajectory of the asteroid Dimorphos in 2022, marking the first time humanity intentionally changed the motion of a celestial object in space and proving that an asteroid's orbit can be altered by kinetic impactor technology.[7]

The key difference between NASA's DART approach and what Industrial DevOps is advocating is that SIL and HIL are combined with a CI/CD pipeline. This allows teams to develop, test, and build cyber-physical systems better and faster.

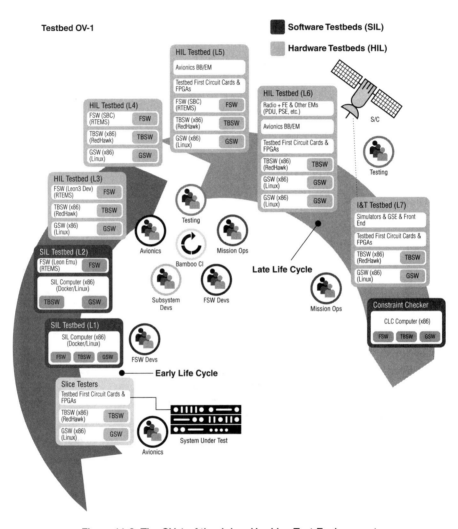

Figure 11.6: The OV-1 of the Johns Hopkins Test Environment

Testing Real-Time Performance

Cyber-physical systems have unique constraints associated with latency, throughput, reliability, and synchronization needs. Fundamentally, latency refs to the delay between the time we input a command and the time the system responds, which has a significant impact. If there is too much latency, the system may miss critical deadlines, fail to respond in a timely manner, or produce incorrect or unreliable analysis and decision-making. To test latency in cyber-physical systems, testers must ensure that the system is optimized for real-time performance by optimizing the system's processors, memory, and network bandwidth under various conditions and scenarios.

Cyber-physical systems are very sensitive to throughput, which is the rate data is transmitted in the system. High throughput is required to ensure that the system can process and respond to data in real time. Throughput will affect data processing, which could result in errors or delays in communication between sensors and control systems, as well as how all the system components work together as a whole. Reliability is also a must when testing cyber-physical systems; even a small margin of error can result in unexpected events and failures. Cyber-physical systems need all the components to operate in sync seamlessly, and if they do not, systems can experience errors in the control systems, such as in a robotic system, which could have unintended movements. Unintended movements are a safety hazard. Consider what an unintended movement could mean in an autonomous vehicle and the potential impact to safety.

Testing for Safety/Security

While not all cyber-physical systems have safety or security concerns, the majority do. Because they are in physical spaces, it's important that we consider the real-world consequences of system failures, errors, or vulnerabilities and the potential impact to human life. A failure in cyber-physical systems such as an automobile or a pacemaker can result in fatalities. While each domain is different, all of them have regulatory requirements that guide system certifications and approvals.

For example, the automotive industry follows ISO 26262, which is an international standard for the functional safety of road vehicles.[8] The standard was first published in 2011 and is now widely used in the automotive industry to ensure the safety of electronic and electrical systems in vehicles. A failure or security breach in a cyber-physical system can have serious consequences for people, property, or the environment, so testing must be rigorous and comprehensive to ensure that the system is safe and secure for its intended use. These testing requirements are integrated early in the development phase and are tested early and often. The challenge is exacerbated in manufacturing because we are pushing out the solution in mass quantities.

Address Challenges with Data

Data has been a challenge in testing systems because often we don't have access to the data, we don't have enough data, or the data that we have is not of sufficient quality. This is a consistent problem for all testing, but it is especially difficult in cyber-physical systems because of the sheer amount of data needed to test through the variety of scenarios.

Automobiles are basically computers on wheels, with multiple sensors to measure anything from engine to wheel speed and everything in between. It was reported by Ford that connected car sensors generate twenty-five gigabytes of data per hour and that autonomous vehicles generate upward of four terabytes in ninety minutes.[9] Data storage and management are difficult. In addition, processing that data in real time requires specialized tools for testing teams to be able to analyze. Data quality is a large concern for testing automobiles. If data is incomplete, inaccurate, or unreliable, the validity of testing results is in question. Data cannot be treated as a side concern; teams need a complete data strategy in order to succeed.

Shift-Left Manufacturing

Just as we want a regular feedback loop from operations into product development, we need a tight feedback loop from manufacturing as we engineer and design cyber-physical systems. Feedback from manufacturing on potential design flaws could be added to our testing approaches. Manufacturing feedback helps optimize test processes by identifying areas that are time-consuming or difficult to test. For example, if a particular component is difficult to assemble or test, it may indicate that changes are needed to the manufacturing process as well.

Approaches to Testing

Testing is more of an art than a science, and there are many approaches to testing to consider as the team creates a testing strategy. We are going to focus on a few that we find particularly important to cyber-physical systems, which include test-first, behavior-driven, risk-based, model-driven, and digital twin–driven testing.

Test-First Development

Test-first development (TFD) is also known as test-driven development (TDD). This approach takes a test-first perspective such that developers convert requirements to test cases before they begin coding. Kent Beck shared this concept with industry in 2003. The approach includes five phases.

1. Convert requirement to test.
2. Run test; let it fail.
3. Write code to pass the test.
4. Run test.
5. Refactor/Repeat.

According to Gartner, the benefits of test-driven development (TDD) include higher quality, better design, and reduced cost of ownership.[10]

Behavior-Driven Development

Changing when and how we test is critical to the success of increasingly complex cyber-physical systems and delivering at the speed demanded by our stakeholders. One method to achieve this is through the practice of behavior-driven development (BDD). BDD is a further evolution of TDD that emphasizes collaboration between developers, testers, and other stakeholders. It focuses on describing the behavior of the system and writing tests to ensure that the system behaves as expected.

The key to BDD is the process of reframing requirements as specific tests, which allows teams to remove the interpretation errors often seen when multiple groups need to interpret requirements. In addition, this practice ensures that teams are focused on business and user outcomes. It also gives teams the ability to detect and correct errors earlier in the process, when it is likely to be more cost-effective and less time-consuming to fix.

Behavior-driven development evolved from TDD in software. It incorporates test-driven development techniques with domain-driven design and object-oriented patterns of architecture (see Appendix B). According to *Forbes*, the benefits of applying BDD to products include improved user experience, faster time to market, and improved product quality.[11]

Behavior-driven development for cyber-physical systems includes the following steps, as shown in Figure 11.7:

1. **Illustrate:** Define the behavior of the system and build a shared understand of system behavior.
2. **Formulate:** Write scenarios/vignettes to describe the system.
3. **Automate:** Write acceptance criteria. Implement the feature. Implement a test and automate the test.
4. **Validate:** Collaborate and get feedback that the system is working as intended.
5. **Demonstrate:** Provide regular, iterative, and integrated demonstrations. Refactor/repeat.

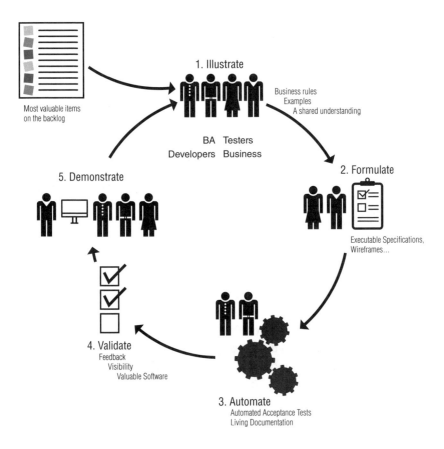

Figure 11.7: Behavior-Driven Development

Let's come back to the CubeSat example. Instead of building a traditional *shall* statement, we can step back and model the behavior of the system, such as a desired behavior of maintaining the CubeSat's attitude for a range of angles. Then we add acceptance criteria that when the CubeSat experiences abnormal sensor behavior, the attitude control system will transition to detumble mode and the CubeSat will maintain an attitude within a predefined range.

Risk-Based Testing

Many cyber-physical systems are safety critical, where system failures or errors have serious consequences. For example, many of the risks in aerospace include sensor failures, communication outages, or system malfunctions. By prioritizing testing efforts in these areas, teams can simulate the catastrophic types of failures and increase the resiliency of systems.

Risk-based testing prioritizes testing efforts based on the level of risk associated with different features or components of a system. Teams start by identifying potential risks associated with the requirements. Then they build tests to mitigate the risks and reevaluate the risk associated with each requirement. No team has endless resources, so risk-based testing allows us to focus on areas with the largest risk score.

For example, for one statement of work a team received, they took the requirements and placed them into an Excel spreadsheet. In one column, they listed the requirement, and in the next column, they listed how they would test the requirement and the associated risk. This simple exercise demonstrated that 32% of the requirements that seemed perfectly acceptable at first look were untestable. Normally, the team would not have found these requirements were untestable until much later in the cycle.

To implement a risk-based testing approach, teams should evaluate systems and requirements for risks with an initial risk score based on information such as past performance or the impact of a system failure. Define the tests for the system areas with goals for mitigating the risks with associated specialized tests. Build or update the solutions to meet the test requirements. After the system is updated, execute the tests to determine where failures occur. Continuously refine the tests in concert with system changes.

Model-Based Testing

Model-based testing is a test approach that uses models to design, generate, and execute test cases. Models are created with Systems Modeling

Language (SysML) and represent the system requirements, design, or behavior. The ideal state would be that test cases are autogenerated from the model with an algorithm that ensures a wide variety of scenarios are considered. Based on our experience, there is still some work in getting tools to do this effectively, but we have hope for the future. The test cases are executed against the system under test to verify behavior. Model-based testing, illustrated in Figure 11.8, supports the complexity associated with cyber-physical systems to ensure the system is thoroughly tested while not being as time intensive as traditional testing due to the automated test generation.

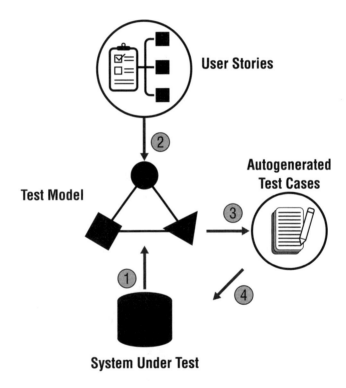

Figure 11.8: Model-Based Testing

The complexity of a single CubeSat is enough to warrant a full suite of models, as illustrated in Figure 11.9, in order to capture functionality of the system and build tests. When we move to a constellation of two hundred or more CubeSats, it becomes even more critical to model and simulate the entire constellation.

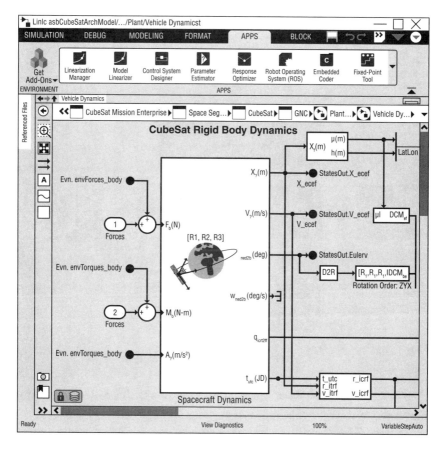

Figure 11.9: MATLAB Model

Digital Twin–Based Testing

Both model-based and twin-based testing utilize models. Digital twin–based testing provides additional fidelity by leveraging a virtual replica of the physical system to simulate the behavior of the physical system and test its performance. Digital twin technology enables the creation of a virtual model of a system that can be used to predict how the physical system will perform in different scenarios and under different conditions. A digital twin uses data from connected sensors to tell the story of an asset by using asset-specific indicators.

Digital twin–based testing, illustrated in Figure 11.10, provides improved accuracy, enhanced safety, and greater flexibility than other testing methods. NASA used a digital twin to test the behavior of the Mars rover's wheels and suspension system. By using digital twins, NASA was able to simulate the rover's behavior in a range of conditions and identify potential issues before the rover was deployed to Mars.[12]

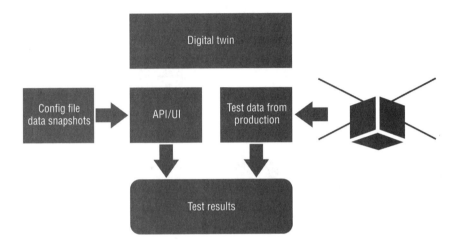

Figure 11.10: Digital Twin–Based Testing

As we've spoken about throughout this book, the benefits of using digital twins to test cyber-physical systems include a reduction in cost by minimizing rework and finding problems before we "bend metal," much faster feedback (digital twins can perform many simulations faster than physical systems can be tested), higher accuracy (because we can analyze data that is difficult to detect in the physical environment), reduced risk exposure earlier in development, and improved communication and transparency across the team.

Multiple Tiers of Testing

Cyber-physical systems require multiple tiers of testing to validate and verify the system. Foundational levels of testing are unit tests, integration tests, systems tests, and finally acceptance tests (see Figure 11.11). We will outline each of these levels and the frequency at which they are performed in the following section. It is important to remember that each of these levels builds on previous tiers. Additional levels of testing include, security, performance, and more.

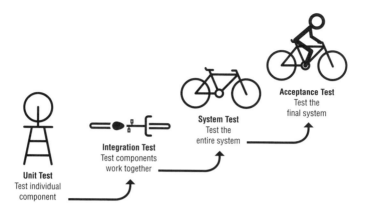

Figure 11.11: Multiple Tiers of Testing

Unit Testing

Unit testing is a technique that involves testing individual units of a system in isolation from the rest of the system. This approach is typically used in software systems, and we consider it table stakes for quality. Unit tests focus on testing small, independent pieces of code, which is crucial because bad code does not scale. Unit tests verify whether the component meets the acceptance criteria or not and greatly speeds up the software life cycle.

In some teams, you may see push back because they believe that it takes longer than just writing the code; however, the benefits are seen in debugging and reduced rework. Important techniques at this level are writing the test that defines the behavior of the component first and then building the capability. Unit testing and TDD are important techniques for ensuring the quality and reliability of systems. By testing individual units in isolation, we identify defects or bugs early in the development process. Unique considerations for cyber-physical systems are the real-time constraints of the system.

Integration Testing

Integration testing is the next level of test after unit testing and ensures combinations of the components of the system work together. Integration testing quickly identifies problems with the component interfaces. There are different approaches to integration testing, such as top-down testing, bottom-up testing, and a hybrid of the two. Top-down testing begins by testing the higher-level components first, then testing the lower-level components. Bottom-up is as it sounds, where we test lower-level components and then build up. Most organizations follow a hybrid approach to integration tests using a combination of both. Top-down identifies early design flaws, whereas bottom-up allows for more parallel development.

System Testing

System testing involves testing the entire system to assess it against the end-to-end system specifications. System testing is performed after a completely integrated system is completed. For cyber-physical systems, this involves testing all the software components, hardware components, and the interactions between them. During system testing, the cyber-physical system is tested under a wide variety of operating conditions, such as normal operating conditions, abnormal operating conditions, and failure scenarios.

Because of the safety-critical nature of most cyber-physical systems—such as in medical devices, aerospace systems, and autonomous vehicles—cyber-physical systems must consider the interactions between the software and hardware components, real-time constraints, safety criticality, environmental variability, and interoperability with other systems. System testing has multiple facets (see Table 11.1), including usability, functionality, performance, security, and compatibility.

Table 11.1: Facets of System Testing

	Test Type	Description	Example
1	Usability	Testing applied that ensures it is user friendly and easy to use, and that it meets the needs and expectations of the end users.	Validating that the displays on an automobile are easy to read.
2	Functional	Testing focused on verifying the system performs as intended and meets the specified need.	Verifying that a spacecraft can communicate with the ground station.

Table 11.1: Facets of System Testing cont.

	Test Type	Description	Example
3	Performance	Testing the system's performance under different workloads and usage scenarios to ensure that it can handle the expected load.	Performing vibration testing to ensure the spacecraft can withstand the vibrations and shock it will experience during launch.
4	Security	Testing that validates the system's security features and ensuring that it is protected against potential security threats and vulnerabilities.	Performing penetration testing that involves simulating an attack on a satellite to determine vulnerability.
5	Compatibility	Testing the system's compatibility with different operating systems, hardware platforms, and other applications.	Testing to validate satellite's communications protocols are compatible with those used by ground systems and they can receive and transmit data

Satellites pose many unique testing challenges, such as latency, packet loss, link flapping, and bandwidth asymmetry; an example of a system under test in this scenario is illustrated in Figure 11.12.[13]

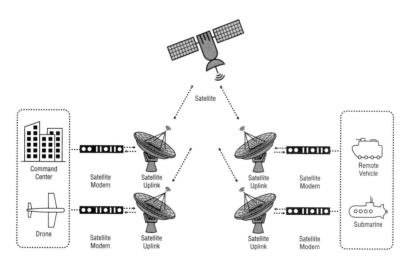

Figure 11.12: System Testing

Chaos Engineering for Testing

Emergent behavior is a growing problem for safety-critical systems, such as satellites or automobiles, as a result of increasing complexity. Emergent behavior is the unpredictable behavior of a system that is based not on an individual component but on the relationships between components. This is like the butterfly effect,* where a small change in one part of a complex system can have significant and unpredictable effects on other components of the system.

The practice of chaos engineering was pioneered at Netflix to test their platforms but has grown in other industries. Chaos engineering can help teams find solutions to these emergent behaviors before they happen, using tactics such as fault injecting. By injecting faults and errors into the system, we can watch how the system responds. Chaos engineering expands upon fault injection to inject a variety of chaos into the system to see how it behaves under stress (Figure 11.13). The goal is to identify issues and build resiliency into the system before it is placed in the real world. The more we intentionally introduce controlled disruptions and failures into the system, the more likely we can identify that emergent behavior and fix it in the system before we have serious consequences.

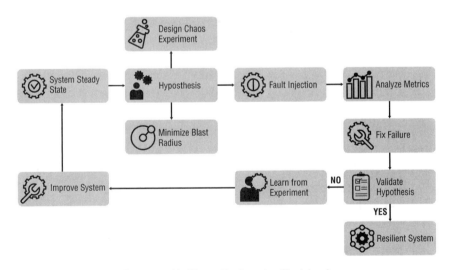

Figure 11.13: Chaos Engineering Explained

---

* The butterfly effect is a known story that says that a flap of a butterfly's wings can cause a tornado to occur in another part of the world.

Acceptance Testing

Acceptance testing comes in many different formats and flavors. If you have been in the Agile community long enough, you are likely familiar with the term acceptance criteria. The acceptance criteria are what is demonstrated during the integrated system reviews. Providing acceptance criteria at the story level helps the development team understand the scope of what they are building, and then they use that information as a basis for discussion as to how they will build and demonstrate it.

The same process is followed at the feature and epic levels. This helps manage scope while defining the validation of what has been developed to ensure the right thing is being built. Stakeholder feedback is provided as the acceptance criteria are defined and during the demonstrations.

For cyber-physical systems, not only do we have acceptance criteria, but we also must complete on-going and final *acceptance testing* against the requirements of the system to ensure safety, security, and compliance standards are met. Early testing may be demonstrated in virtual environments, but the final acceptance testing is completed in the hardware-in-the-loop environment, reflecting as close as possible the operational environment.

CASE STUDY: Idaho National Laboratory Digital Twin–Driven Testing Could Enhance Safety[14]

Nuclear reactors are the epitome of large, safety-critical, cyber-physical systems. Not only are these systems complex, there are also significant challenges when it comes to testing, such as requiring expensive, specialized equipment such as radiation detectors and ultrasonic equipment. Many of the tests can take months to complete.

To address this problem, in 2022, the Idaho National Laboratory built a virtual model of the Microreactor Agile Non-Nuclear Experimental Testbed (MAGNET) using sensor data and open-source technologies. They integrated machine learning to evaluate and test microreactor behaviors under different operating conditions. The researchers were able to detect trends in unfavorable threshold temperatures. The digital twin autonomously adjusted the temperature in the heat pipe to avoid potential future complications.

They believe the success of the experiment could revolutionize the nuclear industry. The autonomous control could significantly reduce operational costs and increase safety while meeting regulatory requirements.

GETTING STARTED

- Adjust your approach to the Vee model by beginning with executable requirements through testing.
- Invest in software-in-the-loop, hardware-in-the-loop, and digital twin labs.
- Develop an intentional automation strategy.
- Incorporate new approaches such as model-driven and digital twin–driven testing and know how you will use them to improve the speed of delivery and quality of the solution.
- Invest in chaos engineering for resilience.
- Define a test-data strategy to shift testing practices left and educate and train people so they know how to use the tools and the associated processes.
- Define a manufacturing-friendly test approach with regular feedback loops to improve verification in the design and reduce rework further downstream.

KEY TAKEAWAYS

- The Vee model needs to be adjusted. Implement the practices defined in the Vee model using an Agile life cycle and shift-left mindset.
- Invest in labs early and build iteratively.
- Automate everything you do more than three times.
- New approaches to testing are required based on the growing complexity of systems.
- Chaos engineering supports resilient systems.

QUESTIONS FOR YOUR TEAM TO ANSWER

- How does a shift-left approach reduce risk to the development of the system?
- What complexities exist for the development of cyber-physical systems? How might you address those challenges?
- Which approach to testing should your team work to improve? Why?
- How does test-driven development change when you are working outside of software?
- What are the benefits of test-driven development?
- What are the challenges of test-driven development?

- What levels of automation are you using?
- What impacts does your test approach have on manufacturing?
- Are there opportunities to increase the fidelity of test approaches by interviewing manufacturing?

COACHING TIPS

- Architect a test strategy inclusive of labs, data needs, and automation.
- Instrument your systems to provide observability.
- Carefully consider safety, security, and regulatory requirements for your system. How will you shift left and address those requirements early? Do you have the talent on your team for those specialty areas?
- Test approaches need to iteratively and incrementally be improved; add a regular cadence to review, and add updates to the product backlog.
- Test automation requires a different set of skills than traditional testing: provide education to upskill your teams.
- Obtain feedback from your manufacturing teams to improve product testing.

CHAPTER 12

APPLY A GROWTH MINDSET

> **PRINCIPLE 9:** Applying a growth mindset expresses the need to continuously learn, innovate, and adapt to the changes around us in order to stay competitive.

Humans continuously evolve and adapt to the changes around us. We do this to survive in an ever-evolving world. A lesson early in my career came when I (Suzette) was working for a startup during the internet boom in the latter part of the 1990s. I received an email from my manager, who was writing about the changes coming to the department, and the message ended with "the only constant is change." No truer words have been spoken. To survive in a world of change means to be continuously learning, inspecting, adapting, and embracing the change around us. Sometimes we are the innovators of the change, the game changers. To continuously learn and grow, we must approach opportunities and obstacles with a growth mindset, realizing we are in a constant state of evolution. We are never done. Change may seem risky to some; however, one of my customers once explained to me that *to not change is the greater risk.*

A Growth Mindset

A growth mindset is best described by Carol Dweck as "the belief that your basic qualities are things you can cultivate through your efforts."[1] It is the ability each of us has that enables us to continuously grow our behavior, skills, performance, talents, or thinking. Through applied learning and resilience, we have seen those who may have felt defeated rise to unimagined success. They explore, innovate, and recreate. They are resilient! A learning organization applies the same growth mindset.

For organizations to thrive and remain competitive means they must also learn and evolve. Organizations that continue to thrive promote learning through experimentation, exploration, and innovation. Through the work of Ron Westrum, we understand that these learning organizations embrace a generative culture.[2] This is a culture where information flows more freely and where people are motivated and passionate about the mission.

This is also reinforced by Daniel Pink, whose research conveys how knowledge workers are often more intrinsically motivated (as long as people are not underpaid) through a sense of purpose and autonomy.[3] Therefore, leaders must communicate the importance of the organization's mission and create environments that support autonomy, mastery, and purpose and drive this into practice. As Brian Tracy says, "If you are not constantly improving and upgrading your skills, you are falling behind, like a runner in a race. And there is no time to lose."[4]

Organizations that approach change with a growth mindset see failure as a learning opportunity. Just like Thomas Edison and his hundreds of experimentations and patents through continuous explorations and seeing failure as a learning opportunity toward new and better solutions. Or the Wright brothers, who conducted many experiments with fast learning cycles leading to the first flight success. Or even more recently, Elon Musk and SpaceX. Through his insatiable appetite for learning, Musk continues to demonstrate progress with space technologies, driving toward his vision to establish a human presence on Mars. These behaviors are reinforced by Eric Ries's Lean Startup model, which values experimentation and learning; customer feedback loops; and short, iterative development cycles to drive toward a business objective.[5]

The Industrial DevOps principle of applying a growth mindset connects to several other Industrial DevOps principles:

- Principle 5: Iterative and fast feedback helps us to continuously learn if what the team is building is meeting the users' expectations. It is essential to fail fast, fail early, learn, and recover quickly.
- Principle 3: Based on objective evidence (data-driven development) of work completed, we can see and measure results with regular feedback from stakeholders.
- Principle 5: The team then applies the learning to the next horizon of planning and adjusts the next iteration of planning and learning.

Having a growth mindset requires a learning culture and an environment of continuous improvement. When working at scale, it is not uncommon for different parts of the organization to have different cultures and social norms. Some areas will have greater tolerance and support for experimentation and more user engagement, while other parts of the organization may have a more top-down or siloed organizational culture. This group culture is defined as "a pattern of shared basic assumptions that was learned by a group as it solved its problems of external adaptation and internal integration that has worked well enough to be considered valid and, therefore, to be taught to new members as the correct way to perceive, think, and feel in relation to those problems."[6]

As we apply new ways of working, new principles, and new technologies, this creates a shift in the organization's culture. Transformation begins with culture and meeting people where they are. A growth mindset also means extending the willingness to think differently about current ways of working and being willing to take the next steps to instill change in your environment. However, organizations cannot simply change the culture; they must provide new experiences that result in changes in behavior that slowly transform the culture. To apply a growth mindset, it is essential for organizations to place value on learning. This means organizations may need to "rethink and unlearn" current ideas and shift away from the notion of "this is the way we've always done things" to questioning and rethinking existing practices and defining a continuous improvement road map.[7]

Organizations have learned from Edgar Schein the importance of providing clarity around the goals of the organization to drive behavior change:[8]

- "The change goal must be defined concretely in terms of the specific problem you are trying to fix, not as 'culture change.'"
- "Old cultural elements can be destroyed by eliminating the people who 'carry' those elements, but new cultural elements can only be learned if the new behavior leads to success and satisfaction."
- "Culture change is always transformative. Change requires a period of unlearning that is psychologically painful."

Applying a Growth Mindset to Your Organization

1. Tie Performance Incentives to Business Outcomes

According to W. Edwards Deming, incentive and merit pay programs have been known to produce the opposite results the organization is seeking.[9] Too often, these programs encourage complacency, focus on results at nearly any cost, and promote a "don't rock the boat" mentality out of concern or fear that any change may negatively impact results. I (Suzette) was once told a story about an effort where developers were incentivized to build a quality product. To encourage quality, a team was paid for every defect they discovered and corrected. As you are likely already imagining, the number of defects continued to rise, and the developers continued to work them off as they continued to be rewarded. This serves as a simple example of how we get what we incentivize. How employees are evaluated and on what basis they are promoted will drive behavior, and that behavior shapes the culture.

Be careful what you measure. It sends a loud message to employees. If you want to drive different behaviors, then change what you are incentivizing or how you are incentivizing.

While most companies have some variation of a performance management system, it has been recognized that, when it's not managed properly, the result and unintended consequence is reduced intrinsic motivation of the worker. When you remove the incentive, you often remove the motivation for the behavior you were seeking, and that behavior becomes greatly diminished unless it has been absorbed as part of the culture.[10] Misplaced or misaligned incentives can unintentionally diminish or destroy innovation and creative ideas.

One of the lessons learned from industry is "claiming victory too soon," which was published by John Kotter.[11] The performance incentive has been removed before the new behavior is ingrained in the culture, and therefore the absorption dies: "Whatever the reason for the effect, however, any incentive or pay-for-performance system tends to make people less enthusiastic about their work and therefore less likely to approach it with a commitment to excellence."[12]

So, what do we do? First, ensure people are paid adequately to remove money from the equation. Daniel Pink also taught us how people are unmotivated if they are underpaid. Pay people well enough.

Second, build a structure that respects and encourages team collaboration, quick learning, and innovation. Ensure compensations are fair and honor collective performance and corporate success. Decide how you will drive and incentivize a culture of collaboration, continuous learning and improvement, innovation, and results-driven behavior.

Third, consider defining team-based goals that connect to the larger organizational mission or product delivery. Build a recognition program where peers can recognize and elevate each other's successes. Build a culture that provides time for innovation and learning, which is an important intrinsic motivator for knowledge workers.[13] Intrinsic motivation supports learning; otherwise, when incentives are removed, so is the learning.

2. Classify Both Successes and Failures as Learning Opportunities.

Amy Edmondson, a professor at Harvard Business School, said, "We are programmed at an early age to think that failure is bad. That belief prevents organizations from effectively learning from their missteps."[14] You may have also heard the mantra "fail fast, fail early." But this thinking is only part of the story. It is not so much about the failure but about the learning and how we use that learning to improve and make better data-driven decisions. In the words of Isao Yoshino, be sure to "celebrate the attempt, not the failure. Failure is the source of so much learning."[15]

When I (Suzette) was working in a tech startup company in the late 1990s, I recall reading an article from a Fortune 50 company. The story was about a project manager who lost a $10 million business effort. I cannot recall the details of the article, but I do recall his fear and his manager's response. The project manager was called into his boss's office later that day, certain he was going to be fired. Instead, his boss handed him his next assignment. The project manager looked puzzled and said, "Thank you. I was sure I was being fired for losing that business effort." His boss replied, "Why would I fire you? We just invested $10 million in your learning."

Take notice. There are different levels and occasions for failure, and not all failure is created equal. With each iteration, we are designing, testing, and integrating. Teams are learning more about the systems they are building, which have become increasingly complex. These short iterations allow teams to quickly pivot, get back on track, and explore options with fast feedback loops. There will be iterations that fail. There will be iterations that are great. Learning is continuous. Improvement is continuous.

Edmondson identifies three broad categories of mistakes where failure happens: "preventable, complexity-related, and intelligent."[16] We want to avoid preventable failures that are found in predictive operations where work is routine and in high-volume production. Complex systems, like our CubeSat mission, take advantage of short iterations, so failures are small and can quickly be corrected and improved upon. This reduces the risk of larger catastrophic failures when it is time to launch. The third area, intelligent, is where new knowledge is acquired. This is where organizations discover, learn, and grow. Learning leads to innovations and can either reduce product risk or build new capabilities, enabling the organization to discover new opportunities to deliver value to customers. The complexity-related and intelligent areas of learning and innovation through small feedback cycles are the target while ensuring the preventable, catastrophic failures are prevented. Be the leader that encourages learning and continuous improvement of products, services, and processes. Avoid the blame game and instead create a culture of psychological safety.

3. Build Psychological Safety

A lack of psychological safety can prove to be a barrier to the adoption of Industrial DevOps principles. Overcoming this barrier relies on the behaviors of leaders in the organization. The behaviors of leaders define whether you have a culture in which people are willing to offer ideas, ask questions, experiment, and raise concerns without fear of penalty, looking bad, or other negative consequences. They encourage employees to speak up, and in turn, employees know their ideas will be respected.

While organizations tout the mantra of "fail fast, fail early, and learn," how do leaders respond when this occurs? The goal of psychological safety is a focus on achieving goals and doing the right thing rather than on self-protection. Psychological safety is an integral component of a generative culture in which there is high collaboration and a foundation of trust permeating the organization. Risks are shared, and failure leads to opportunity and growth. These behaviors must begin at the top, with leaders. Then those leaders demonstrate and coach their teams on these behaviors, who then coach their teams, and so on, resulting in a high-performance organization.

Let's look at a team who was early in their Industrial DevOps journey and had completed a two-week iteration. Things had not gone well. Their customer arrived for the demonstration of completed work, but there was little to nothing to demonstrate. The product owner, concerned with how the customer would react, took the stand and shared what had been planned, the challenges and impediments the team had faced, and their recommendations for moving forward. She explained that it was good the team had found the problems early, and the team would be able to make immediate corrections in the upcoming iterations.

This early discovery had prevented a great schedule slip and potential increased costs if the problems had been found later in the product development cycle. Early integration and testing resulted in new learning and uncovered unknown issues. And importantly, the leader of the team exemplified psychological safety by not blaming her team but espousing the learning that had come from the "failure."

At the conclusion of the product owner's explanation as to why there was nothing to demonstrate (that is, progress was now in the "red"), yet it had resulted in beneficial learning, the customer stood up and said, "So, what you are telling me is that red is the new green?" Even though the project was in the "red," the leader had understood how this was actually a good thing (green). It was understood this was not catastrophic failure but iterative learning that could prevent catastrophic failure in the future.

By creating an environment of psychological safety, people will speak up, share their ideas or concerns, share risks, and share failures, which results in increased innovations, learning, and successes.

4. Invest in Learning and Talent Development

Adopting Industrial DevOps principles for improved efficiency and delivery of value requires learning new principles and practices in all parts of the organization. Creating shared learning experiences shapes a common language and shared mental models across functions and teams, helping to create a single team of teams. The implementation and ongoing evolution of the latest digital tools and smart manufacturing technologies also requires ongoing investment. Organizations must continue to invest in the technical development of employees at all levels as digital capabilities continue to evolve at a rapid pace.

Build your learning strategy aligned with your Industrial DevOps and digital transformation strategy. To begin, understand the current state and experiences of the organization and where the organization is investing in digital capabilities and process improvements. Training and learning in an organization can occur in a myriad of ways, from classroom training (virtual or in-person), flipped classroom learning experiences, to Dojos and on-the-job training. The important thing is, regardless of the format, there must be intentional application.

In *The Six Disciplines of Breakthrough Learning: How to Turn Training and Development into Business Results*, the authors highlight four factors necessary to apply new learning on the job: (1) personal capacity to make changes, (2) the opportunity at work to apply the learning, (3) the teaching of the learning, and (4) the perception of the new learning as relevant and applicable.[17]

Training efforts as part of the learning journey must be designed to address an individual's need for new knowledge and skills. Learning goes through several steps in order for results to be realized. There is the assumption that the drive toward improved results requires new learning and this new learning is one of the gaps between the current state and the desired results (Figure 12.1). The level to which the learning is applied is impacted by several factors, such as the environment, leadership, reinforcements, and early successes or failures as a result of the new learning (as detailed in *The Six Disciplines of Breakthrough Learning*).[18]

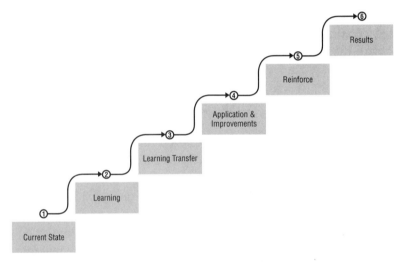

Figure 12.1: Steps of Learning

Leaders play a key role in this process and directly impact the transfer of learning and the climate toward application. In 2008, Fast Company published the article "Change or Die," which highlighted how difficult change is for many people even when their health is at risk.[19] To help people understand why the change needs to happen, leaders must communicate the urgency and need for the new ways of working. They must ensure their behaviors reflect the new model and their words are aligned with their actions.

The application of the learning needs is reinforced through leadership's active participation and communication, coaching, and opportunities to share their experiences driving improved performance. This reinforcement and encouragement by leadership is necessary to keep the momentum and help the new behaviors stick. However, even when people understand why change is needed and they know they have leadership support, going from understanding of new ways of working to implementation is difficult for many teams.

5. Support Coaching and Cross-Domain Learning

The importance of coaching is evident in sports programs. The coach is present and engaged with the team, helping them prepare to go out on the field and score the goal. They inspire, encourage, and help shape a path forward for team execution. Coaching is also important for both Lean and Agile teams. A report from the US Government Accountability Office noted in their 2020 assessment that "one of the biggest obstacles to an Agile transformation can be that very few people in the organization know and understand Agile methods or that they implement Agile based on limited experience and understanding of them. As a result, sponsors and senior stakeholders may need training and/or coaching regarding their new responsibilities."[20]

Without a coach supporting and encouraging the adoption of consistent practices and commitment to continuous improvement, the efficiency and effective delivery of value is greatly diminished. This has been observed through multiple pathfinder programs within US federal government studies. Further research continues to demonstrate the value in having coaches help the organization and teams in their adoption. They help the team and organization improve their practices to deliver value with measurable results while building high-performing teams. The coach helps the team and leaders fulfill their roles in the process.[21]

According to a recent *Forbes* report, "Why T-Shaped Teams Are the Future of Work," the importance of T-shaped skills is on the rise.[22] Organizations continue to face significant gaps in the skills they need at a time where digital capabilities are growing exponentially. For teams to deliver faster and fill gaps, T-shaped skills and cross-domain learning are critical. T-shaped skills mean an individual has depth in one or more specific areas and breadth in a wide variety of areas. Individuals are never as smart as the team.

Coming back to our CubeSat example, we may have a person who understands how to make updates to the attitude control system; however, when working in collaboration with the broader team, they may also have experience in cybersecurity controls or a better approach to test automation.

An early study in the biotech industry reports on the importance of T-shaped skills.[23] The finding indicates "that under the high technology uncertainty, the effect of T-shaped skills on innovation speed is significantly strong." Not only have we witnessed this growing need in organizations, but research is beginning to validate these observations. Organizations need a talent-development strategy that includes the cross-skill training to upskill our talent, enabling the team to respond faster and reduce hand-offs between functions.

6. Leaders Lead Best by Example

For these techniques to be implemented within the organization requires the leadership teams to be actively engaged and modeling the new behaviors. Leadership and organizational culture are intertwined. As Schein says, leaders create, change, and sustain organizational culture.[24] Only they have the authority and monetary power to change the existing operating system in the organization. Therefore, leaders are ultimately responsible for driving the necessary culture for learning, experimentation, and innovation.

How leaders behave, how they respond, and the words they choose all impact the behaviors of others. People watch and they listen. They follow the leader's example: "All the fine speeches, the lovely value statements, and the well-intended training will have little effect or could even have a negative effect if what is said and what is done do not match."[25]

One of the key enablers for Industrial DevOps to be successful is for leaders to be educated in the techniques. During the midst of the transformation, I (Robin) found it interesting that every leader inside of my company was bragging to our customers about our experience and commitment to Agile and DevOps.

While this may have been accurate at some level of the organization, all but one of the executive leaders had not made time for training, explaining they were too busy. In some cases, they may have offered up a spare hour on their calendar. Their claim of commitment did not match their actions. Their behavior was witnessed by their employees, who, in turn, realized their claims of commitment were disingenuous. The importance of authenticity in leadership has been well documented—that is, leaders model the way through their words and actions. Culture is anchored in "unspoken behaviors, mindsets, and social patterns."[26]

Jack Welch, former CEO of GE, demonstrated "the courage to be open; to welcome change and new ideas regardless of their source. . . . real self-confidence is not reflected in a title, an expensive suit, a fancy car, or a series of acquisitions. It is reflected in your mindset; your readiness to grow."[27]

Harvard Business Review's "The Leader's Guide to Corporate Culture" offers a framework leaders can use to assess the style of the organization's culture. Assessing the culture provides input as leaders strive to align strategy toward a new desired state. The authors define four practices that leaders can do to support successful culture change.[28]

1. Articulate the aspiration and the desired outcomes.
2. Select and develop leaders who align with the target culture. Provide training for leaders to help understand the benefits of the desired change. Be prepared that some people will move on feeling they are not a good fit in the new ways of working.
3. Use organizational conversations. Leaders not only communicate the *what* and *why* of the new ways of working but also demonstrate it. They actively engage in planning events or demonstrations. They talk about innovations and demonstrations of working products.
4. Reinforce the desired change through organizational design.

Leaders can create a road map to focus on the behaviors that shape the desired new ways of working. We recommend starting with four improvement areas: mindset validation, support structures of the organization, technical competency, and role-modeling behaviors, starting with leadership.[29] Using a quarterly road map, the organization defines a path forward in each of the improvement areas, with each quarter resulting in some tangible, demonstrable outcome.

Conclusion

The road map for change will evolve as the organizational system matures. Leaders must focus on building awareness and sharing case studies that demonstrate what change looks like. As they begin to see positive results, they should capture these successes and share them across the organization. Success builds more success. Leaders should also ensure alignment with their promotion system to include team-based awards.

Building a learning culture with regular opportunities to grow is a key component to the successful adoption of Industrial DevOps. It requires new skills and competencies. Leaders must demonstrate their commitment to understand and model the principles, build environments of psychological safety for innovation and fast learning, actively participate in removing organizational barriers, and show a commitment to continuous learning and relentless improvement.

In 1984, a new joint venture experiment was started by Toyota Motor Corp. and General Motors (GM). The GM manufacturing plant was located in Fremont, California. At that time, the culture at the GM plant was described as highly dysfunctional. The quality of the products seemed insurmountable, and absenteeism continued to climb and reached as much as 20%. It was obvious something had to be done.

At the same time, Toyota was interested in doing business within the US but needed to learn more about unions and how to participate in the US labor market. With Toyota's strong Lean manufacturing practices and the Toyota Production System, which produced high-quality products and a culture of continuous improvement, the joint venture was ideal. It was seemingly a win for both parties. Toyota was very much focused on learning and improving quality, and this was something GM Fremont was in great need of at that time. This joint venture became known as New United Motor Manufacturing, Inc. (NUMMI).

According to John Shook, all those in a supervisory role (to the team level) were sent to Toyota for a minimum of two weeks to learn the Toyota Production System, a system of tools and practices built on Lean manufacturing principles.

Over time, improvements began to emerge at Fremont. As a result of embracing Lean thinking and applying their learning, the Fremont plant went from its completely dysfunctional state into a *model manufacturing plant*. At the start of the joint venture, GM Fremont had the worst quality. Now they were demonstrating the highest quality. Employee morale improved and absenteeism dropped from 20% to 2%. How did the culture change? Well, culture is the commonly held set of beliefs, behaviors, and thinking of a community, and they started with "doing." Leadership started by changing "what people do rather than how they think."[31]

According to the article published by Shook, the foundation that created this turnaround was clarity on each employee's job through standardized work, followed by the training and resources needed to perform that job. They adopted a problem-solving, continuous-improvement mindset and were provided specific tools that they could apply. At the conclusion of the training, participants were asked what they needed most to move forward and apply their new learning. While stated in different ways, what people wanted was the ability to address the challenging problems they faced every day without fear of blame.

GETTING STARTED

- Review your current incentive structure to ensure its tied to business outcomes.

- Survey your teams to measure psychological safety.
- Assess your culture intentionally architect your culture just as you do with your system.
- Invest in being a learning organization.
- Train leaders to be coaches for your culture of the future.
- Challenge your teams to identify ways to further improve culture.

KEY TAKEAWAYS

- Create shared language and mental models.
- Create a learning culture.
- Leaders shape the culture.
- Build psychological safety and remove the blame game.
- Leaders must demonstrate a growth mindset and embrace training.
- Leaders must be authentic.
- Behavior change leads to culture change.
- Build shared mental models and understand different perspectives.
- Use a framework to understand your culture.
- Articulate the desired future state and the benefits.
- Create a road map for change.

QUESTIONS TO ASK YOUR TEAM

- How is your leadership team demonstrating authenticity, empathy, and psychological safety?
- What new learning has your leadership team actively engaged in? How have you shared that with the organization?
- How are you incentivizing or reinforcing the new behaviors?
- Have you built a learning strategy that aligns with your Industrial DevOps strategy?

COACHING TIPS

- To change culture, you need to provide new experiences and the resources necessary for people to do their job.
- Invest in the technical development of employees at all levels as digital capabilities continue to evolve.
- Consider what you measure.
- Ensure compensations are fair and honor collective performance and corporate success.

- Create a road map to focus on the behaviors that shape the desired new ways of working.
- Build your learning strategy aligned with your Industrial DevOps strategy. To begin, understand the current state and experiences of the organization and where the organization is investing in digital capabilities. Training and learning in an organization can occur in a myriad of ways, from classroom training to Dojos to on-the-job training. The important thing is, regardless of the format, there must be intentional application.
- Discuss with your team what it means to have a psychologically safe environment and what behaviors or words create this environment.

PART III
FORGING THE PATH FORWARD

CHAPTER 13

BRINGING IT ALL TOGETHER

The principles of Industrial DevOps come together to build better systems faster. Industrial DevOps enables discovery, learning, and value delivery to the customer. Getting started on the Industrial DevOps journey for large systems development can seem like a daunting undertaking. To help your organization along this journey, we have outlined a framework to pull together the Industrial DevOps principles along with organizational enablers to support a successful transformation.

Each section of the framework (see Figure 13.1) is outlined in the sections that follow and includes a description, why it is important, and one or more actions you can take as you strive to "bring it all together." Based on our research and experiences, these are success patterns of many organizations; however, as organizational culture and needs vary, so will your implementation road map and the results. Take time to understand your organization's culture, engage others, and build improvements from there.

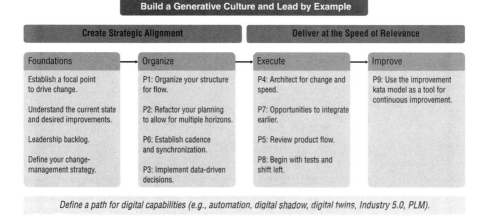

Figure 13.1: Industrial DevOps Framework

Industrial DevOps Framework

STEP 1: Build a Generative Culture and Lead by Example

As organizations transition into a digital and data-driven era, they must also shift their ways of working and build a performance-oriented organization. Whereas bureaucratic organizations once ruled during the Industrial Age, leadership is learning that a more generative culture drives better results today (see Table 13.1).

Bureaucratic organizations experience a top-down heavy hierarchical structure, with command-and-control leadership styles, where people are told what to do and have little voice in the decision-making process. This culture is more likely to encourage blind obedience to strict standards that leave little room for flexibility or autonomy at the team level, resulting in limited commitment to a continuous improvement mindset and limited results.

Table 13.1: Westrum's Organizational Typology Model

CATEGORY	DESCRIPTION	CHARACTERISTICS
1. Pathological (Power oriented)	Organizations are managed through fear and threats. People are incentivized to hoard or withhold information to improve their power stance.	Low cooperation Messengers shot Responsibilities shirked Bridging discouraged Failure leads to scapegoating Novelty crushed
2. Bureaucratic (Rule oriented)	Organizations protect departments. The members of the department want to lead the organization and follow a strict set of rules where all members are treated equally.	Modest cooperation Messengers neglected Narrow responsibilities Bridging tolerated Failure leads to justice Novelty leads to problems
3. Generative (Performance oriented)	Organizations focus on the mission. The organization implements the mission by intent. Everything is about successfully meeting goals and objectives.	High cooperation Messengers trained Risks shared Bridging encouraged Failure leads to inquiry Novelty implemented

Organizations recognize that in order to survive in the digital era, they need to allow time for innovation and greater autonomy at the level of the organization where work is performed. This is the foundation of a generative culture, in which the organization is aligned around shared goals, a mindset of shared success, and the belief that we learn through failure, leading us to bigger and better innovations. This results in a workforce that embodies continuous learning and experimentation, knowledge sharing, growth, and value to the customer. It is the culture of the future and the culture of organizations that not only survive but flourish throughout this transformational period.

The culture of an organization is defined and reinforced by the behaviors of its leaders. People follow the leader. According to "The New Leadership Playbook" from *MIT Sloan Management Review*,

> . . . the principles of behavior, culture, and organizational design that once defined excellence have become less relevant and less valuable. Self-deceptive inconsistencies and contradictions dominate the self-assessments of many leaders today.[5]

By applying a growth mindset, your leadership teams recognize the need for continuous improvement and learning. Be intentional in building communities of leaders at all levels of the organization; find your "strategic, cultural, human-capital, and personal blind spots"; and add to your road map items that will shape the organization toward a generative culture.

STEP 2: Strategic Alignment and Deliver at the Speed of Relevance

The goal is not Agile or Lean or DevOps or digital transformation. The goal is to build better systems faster to deliver value to the customer. We believe that applying the principles of Industrial DevOps to the building of cyber-physical systems is the way to achieving that goal. Organizations identify their strategic goals and then align teams around a set of defined business outcomes specific to their strategic goals. This builds a surge of alignment across the organization and helps the organization to focus on what is important.

There are a variety of approaches organizations can take to define and align their strategy. Two popular approaches are Hoshin Kanri, from the

Lean community, and objectives and key results (OKRs), often used by software companies.

Hoshin Kanri is a well-defined planning process used by many organizations that have a Lean background. Leadership begins the Hoshin Kanri process by defining a vision and high-level objectives that align with the needs of the customer. These objectives are then communicated and refined using a series of feedback loops with different parts of the organization. It traverses between upper management, middle management, and their team members as objectives are formulated. This approach includes engagement from many areas and strives to ensure all levels of the organization have had the opportunity to have a voice in the process.[6]

Another popular approach is aligning strategy to OKRs. This approach was made popular by John Doerr of Google in his book *Measure What Matters*. The objectives describe what it is you want to do and by when. The key results are how you will progress toward your objective. An objective typically has three to five key results. They are specific and time bound. Once the key results have been completed, it is assumed the objective has been achieved.[7]

Both approaches are well understood in industry and are designed to help organizations align around the prioritized business outcomes and provide the direction forward. Leaders define *what*. Teams are empowered to define *how*.

STEP 3: Build the Foundation

Once you have determined that Industrial DevOps will add value to your organization, you will need to build a foundation for your journey. Change at scale requires a leadership focal point, situational awareness, and change management. Below are several actions your organization can take to get started.

Establish a Focal Point to Lead Change

Establish a focal point to lead the change for improving speed and agility. According to the authors of *From PMO to VMO*, "Achieving true end-to-end business agility requires transitioning to this new Agile VMO [Value Management Office] structure, as well as the methodical restructuring of processes and structures across the entire value stream, from business strategy to operations and every step along the way."[1] The organization has the responsibility to build value streams, define the minimum standards for quality and governance, improve and measure the efficiencies, and build the enablers for continuous improvement and innovation. This part of the organization may find itself to be the focal point, the guiding coalition, for the change initiative.

In the book *Applied Operational Excellence for the Oil, Gas, and Process Industries*, the authors explain that organizations often apply operational excellence "through a comprehensive, structured, and well-organized approach to building organizational capability as well as to leverage benefits resulting from doing things the right way."[2] It includes the standards, safety requirements, and other quality enablers to ensure safe environments for workers, which is a criterion for the building of most cyber-physical systems. Therefore, if you already have an organization focused on operational excellence, they may be the ones who choose to take on ownership of the transformation. These are only two examples of who might lead the change in your organization.

Whether through your VMO, operational excellence, or another part of the organization, the important thing is to have a clear understanding of who has the responsibility to be the guiding coalition for leading the transformation.

Understand the Current State and Desired Improvements

One of the initial steps the leadership team takes as a guiding coalition is to gain situational awareness. One of their first steps is to understand and capture the current state of the organization. To establish the current state, a working session is conducted to take stock of the current strengths, weaknesses, opportunities, and threats while identifying the current skills, experiences, and knowledge of those impacted by this new way of working. To gather information, you might conduct a cultural assessment, gather peer/competitive data, capture pain points from different parts of the organization, or perform a benchmarking exercise.

As part of this exercise, discuss and capture the desired future state. Consider what the future will look like after the organization adopts Industrial DevOps principles. Imagine, for a moment, chart paper around a conference room. The team is actively engaged in discussing and capturing what this new future state will look like. Each participant is capturing their ideas for the future state on Post-It® notes. Then they discuss prioritization of the collective input. Next, they perform a gap analysis between the current "as is" state and the desired "to be" state. The data and insights gleaned from this activity help the team visualize the activities that need to be accomplished. They begin to converge on the set of actions needed to move forward.

Transformation Backlog

The leadership team takes their newly found discoveries and begins to build the backlog based on the set of actions that have been prioritized. This backlog will also include the work of the leadership team as they define the steps they want to take to build a generative culture. These steps address key areas such as mindset, the enabling structure, the skills and knowledge needed for Industrial DevOps implementation, and the desired leadership competencies.[3]

Define Your Change Management Approach

Transformative change initiatives require a change management approach to address many needs within the organization: from building awareness of the initiative, creating a sense of urgency, capturing pain points, leveraging existing successes, building technical competencies and enablers, and reinforcing the new desired behaviors. Ensure a solid stakeholder management plan with an approach to communicate progress and gather feedback while continuing to support the needs of the organization. There are several change management models your organization can choose from to get started. Popular models include Kotter's eight-step change model and Prosci's ADKAR (Awareness, Desire, Knowledge, Ability, Reinforcement). It is likely your organization already has a change management framework they prefer to use. Use it. Prioritize and integrate the change management activities as a set of features in your quarterly road map.

STEP 4: Organize

P1: Optimize Your Organizational Structure for Flow

Evaluate your current organizational structure and determine if your structure is optimized to meet your goals. Some organizations are organized for economies of scale, and others can be organized for speed of delivery to the customer. Speed of delivery is the most frequent goal for meeting market demands in the digital age. If that is the case for your organization, then your goal is to inverse Conway's law (see Chapter 7). Inversing Conway's law requires flattening the organizational structure, creating cross-functional teams, and fostering a culture of collaboration. As you flatten, one consideration is to ensure you maintain an organizational structure that supports specialized skills development in areas such as modeling and software development. Creating communities of practice (CoP) or communities of excellence (CoE) is an excellent mechanism to support skills development. Your current organizational structure evolved with your product or service architecture, which means that you must take a stepwise approach to flattening your organization at the same rate as you refactor your architecture or you will run the risk of not being able to manage communications or dependencies effectively. For some organizations, this change may take years, but with each incremental step, you should experience increased flow and speed of delivery.

P2: Refactor Your Planning to Allow for Multiple Horizons and P6: Establish Cadence and Synchronization

Managing expectancies for schedules and timelines is important; however, we have discussed how predictive planning does not provide the flexibility or transparency needed to be successful. It's likely your organization currently has a traditional schedule for your complete product or service delivery. If so, then you will want to ensure it is set up in a way that gives flexibility and visibility into progress and aligns with the work of the teams.

Begin by reviewing the schedule for:

- Structure
- Decoupling scope from time
- Level of detail
- Length of time remaining
- Resource-loading approach
- Business rhythms needed to update with empirical data

Structure

When reviewing the schedule for structure, you are looking for whether it contains large phase gates where you are completing all of a certain type of activity and then holding a milestone review. For example, many times when the DoD contracts for a new satellite, they specify specific gates and culminated with formal reviews. The goal is to provide confidence that the teams fully understand the requirements for the system being developed. The problem is this provides a false sense of security since this is typically only on paper and none of the system has been verified or validated. A schedule that prescribes all of the design is to be completed before implementation often results in extensive rework. There is a need to migrate to a schedule that designs, builds, and tests a component or subsystem with a milestone that identifies components demonstrated and validated.

It may be the case that your current program contract requires phase gates. For a temporary work-around during transition, develop a partner feature schedule, as illustrated in Figure 13.2, that shows component completion as well. This allows you to start getting validated capabilities sooner. Notice models and digital twins are a feature themselves, and they are required to be maintained to get rapid feedback on feature updates.

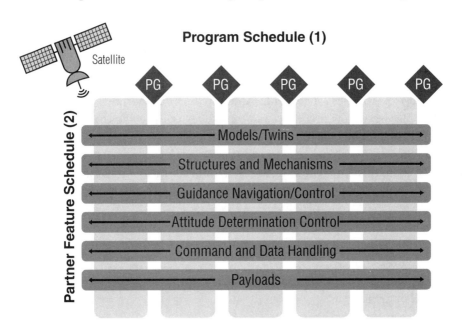

PG = Phase Gate

Figure 13.2: Schedule Grid

Decouple Scope from Time

A typical schedule tightly couples scope and time together. Epics and features are the scope needed to fulfill requirements that have been relatively sized. Time is years, half years, and quarters. We recommend identifying epics and features with key acceptance criteria. Then place the scope, such as epics and features, into time such as quarters and half years. Once we have placed scope in time, then we link predecessor or successor dependencies. The benefit of decoupling scope from time allows us to easily move scope around based on priorities.

Level of Detail in the Schedule

When reviewing the schedule, collapse small tasks into bigger chunks of work We will place the smaller, detailed work in the backlog. If you have a number of these small tasks, we recommend that you migrate those smaller tasks to the product backlog and link them to higher-level tasks within the schedule using a unique identifier. The goal of this is to create flexibility with the smaller tasks while providing traceability and support dependency management at the higher level.

Length of Time Remaining in Schedule

Review the schedule for the total length of time. For a five-year schedule, the goal would be to decompose into the varied time horizons. For year one, decompose the schedule into features planned by quarter; the second year, we recommend decomposing epics planned into half years. For the remaining years, place high-level epics in the out-years. Adjust your approach to the length of the schedule.

Resource Loading Approach

Next, we want to evaluate how the schedule has been resource loaded. In a typical schedule, individual resources are assigned to tasks. One of the goals for Industrial DevOps is team-based collaboration and delivery, which requires a move from individual resource loading to team-based loading. The mechanism of team-based loading is creating a fictional cross-functional team with capacity and cost and assigning teams to the feature-level tasks.

Business Rhythm to Update with Empirical Data

One of the tenets of Industrial DevOps is moving from predictive planning to empirical planning. Given that we need to use empirical data to adjust the plans, we need business rhythms with an established cadence that intentionally reviews data and updates the plans. Typical programs may do this weekly or every other week, depending on the size of the program and the level of volatility.

Maximize Team Alignment

We want to obtain agreement for a cadence that the teams can synchronize on. Our goal is that we can optimize the delivery of the system by using a common cadence. When cadence is implemented in conjunction with iterative and incremental time cycles, we obtain the maximum possible feedback at the system level. Given the potential lead times for hardware, there are many teams who select a longer sprint cycle to accommodate integration of hardware and software items. When we talk about cross-functional teams, we may have hardware and electronics on one team if they are collaborating on a specific set of features or software and hardware on a team if the component requires frequent modifications, but we do not necessarily need every skill type on every team—only the skills associated with a specific product we are working on.

P3: Implement Data-Driven Decisions

Industrial DevOps wants to use quantitative data to measure progress or make decisions. Moving away from phase-gate milestones requires different approaches to obtain and track program progress. Leverage product demonstrations and user feedback to build confidence in your progress on products. The benefit of the demonstrations is that we get a real understanding of where the validated product stands. If we can not demonstrate working capability, we likely do not have a validated capability.

Quantitative data regarding the value of a capability allows us to move away from a gut feeling when we prioritize work. Although many of us trust our gut instincts, there is significant research that shows we are almost always incorrect. Limiting work in progress and ensuring flow requires prioritization of the work so we can collaborate across the team on working products.

STEP 5: Execute

P4: Architect for Change and Speed

As we have discussed throughout the book, to deliver products at speed, we need to have modular architecture with standardized interfaces. If you have a legacy program, you may have a monolithic architecture that is tightly coupled. For Industrial DevOps, to provide optimal results, we recommend using a pattern known in software engineering as the strangler pattern, which is a pattern to gradually modernize legacy systems. We begin by identifying strangulation points, which are areas of the system that we need to replace. We next create integration points to enable communication between old components and new components. As illustrated in Figure 13.3, we incrementally replace components in the new system with new modular components with well-defined interfaces. We need to iterate the system architecture with the organizational structures changes, which may take years, depending on the size and complexity of the system. With each update, we run a complete verification and validation of the change to ensure the system works as expected. If we consider our CubeSat example, we may need to decouple the attitude control system from the guidance, navigation, and control system so that we could move the monolith to microservices. If we were to consider performing something similar with hardware portions, this could be something as simple as creating an adapter for the CubeSat deployer while we try out a new design for the deployer connector.

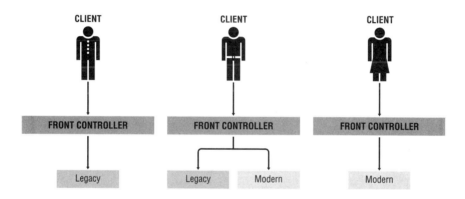

Figure 13.3: Strangler Pattern in Action

P7: Find Opportunities to Integrate Earlier

We discussed the need to integrate as frequently as feasible. Factors to consider when scheduling integration points are system size, complexity, level of automation, available environments, and associated risk. The larger and more complex the system, the more time it takes to integrate updates, which may make it unfeasible, especially if we are starting with a highly coupled system.

To reduce integration time, it will be important to refactor the solution into smaller modules. In addition, if we are beginning with all-manual test approaches, integration takes much longer. Defining an automation strategy early is to reduce the time between integrating. In traditional development, many companies choose not to invest in expensive testing and integration environments until later in the development cycle. As we transition to Industrial DevOps, we need to strategize how to obtain environments as early as possible. Leveraging software-in-the-loop (SIL), hardware-in-the-loop (HIL), and modeling environments allows us to integrate the system faster, at least in the virtual space.

P5: Review Product Flow

Begin by documenting the current system flow. Identify items such as key milestones, and review dates as well as lead time of new features delivered to customers or cycle time to make changes in the system. Also, measure the length of time between feedback loops. Feedback measures could include the frequency of feedback from users, business owners, manufacturing, suppliers, or any other members of the value stream whose feedback helps validate the system. From here, we break the work down into smaller items and put together a cycle from which to iterate.

P8: Begin with Tests and Shift Left

Moving tests earlier in the life cycle reduces risk and lowers rework exponentially. Leveraging BDD approaches enables us to begin with executable requirements. Evaluating how the requirements will be verified and validated before building the system allows us to identify untestable requirements early. A key enabler for testing early is ensuring you have the tools to do the job. If we delay getting modeling and hardware-in-the-loop environments, we will be unable to run tests early. If we delay in developing a multi-tiered test automation strategy, frequent testing is unfeasible. When we integrate early testing with frequent integration, we have the ability to learn and deliver much faster.

STEP 6: Improve

P9: Use the Improvement Kata Model as a Tool for Improvement

The Toyota improvement kata is a problem-solving model, based on scientific experimentation, that individuals and teams use to iterate and improve toward a new, desired future state. Lean expert Mike Rother defines the improvement kata as "a skill-building process to shift our mindset and habits from a natural tendency to jump to conclusions, to a tendency to think and act more scientifically."[4]

How does this fit with the current and future-state activities? As the organization moves toward the future state that it has defined, it will apply an improvement kata approach to continuously and iteratively improve the adoption of the Industrial DevOps principles. Coupled with the improvement kata is the concept of the coaching kata.

The idea is that leaders and managers are the coaches in this process. They lead by example and coach their teams through improvements to develop new skills. We also recommend employing coaches experienced in Industrial DevOps to assist leaders and the teams as they adopt and improve the practices associated with each of the principles. The improvement kata and coaching kata are two well-documented techniques from the Lean community that you can employ as you iterate, learn, and improve toward the next future state.

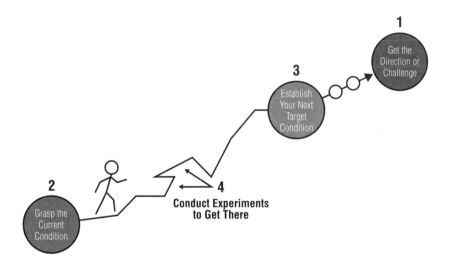

Figure 13.4: The Toyota Improvement Kata
Source: Mike Rother, Toyota Kata.

STEP 7: Define a Path for Digital Capabilities

Industrial DevOps principles consider the integration of people, processes, and tools. Many of these principles leverage digital capabilities that require both budget and investment to be realized. Create your digital capabilities road map based on the priorities and greatest return on investment

(ROI) of your unique cyber-physical system. Ideas for your road map might include things such as:

- Ensure tools leveraged across the value stream are integrated through enabling architecture and APIs. Build a continuous integration and continuous delivery (CI/CD) pipeline early in your project and incrementally enhance the fidelity at each iteration.
- Capture personas, models, and scenarios to define and improve understanding of the system .
- Invest early in labs that contain simulators, emulators, hardware-in-the-loop, and 3D modeling.
- Use your 3D models to build augmented and virtual reality to further visualize system capabilities.
- Build digital shadows, which evolve into digital twins as more of the system is built out.
- Generate digital threads that integrate temporal data to provide increased insight.
- Leverage 3D printing/additive manufacturing to buy down risk earlier or use to build the physical product itself if cost-effective.

Relationships between the Principles of Industrial DevOps

This section outlines the principles of Industrial DevOps and demonstrates how each principle is connected with the other principles. It's important to remember that the principles work together and should be looked at holistically to achieve the greatest benefits.

As you define your Industrial DevOps journey, consider each of the principles and how they are interconnected. For example, if you choose to focus first on shift-left integration and testing but are still working in functional silos versus organizing teams around value delivery, you will not get the results you are anticipating.

We have organized the principles into three primary categories:

1. Organization and Structure
2. Execution
3. Continuous Improvement

Table 13.2: Principles Focused on Organization and Structure

P1: Organize for the Flow of Value

Connection to Other Principles	P2: Apply Multiple Horizons of Planning P4: Architect for Change and Speed P7: Integrate Early and Often
How the Principles Work Together	(P2 and P7) Once the value stream is understood, teams are organized around the value stream. The product is defined from vision and high-level yearly plans, with detailed backlog definition happening closer to execution. At the implementation level, teams are cross-functional and develop the product through a series of short iterations with frequent feedback loops to regularly validate and verify the features while continuously integrating and testing throughout the iterative development cycle. (P4) Consider how the system is architected and how the organization aligns the teams. The teams must be organized around the flow of value delivery. Define your value stream and the products that the value stream produces. Organize the people around the flow of value. Revisit the architecture of the system and ensure modularity and reduce dependencies. While the backlog defines new functionality, it must also include the work that needs to be done to continuously evolve the architecture.

P2: Apply Multiple Horizons of Planning

Connection to Other Principles	P6: Establish Cadence and Synchronization
How the Principles Work Together	Each of the multiple horizons of planning (P2) occurs on a regular cadence. Each horizon of planning yields empirical data demonstrating the success of the plan as it is executed and identifying necessary adjustments. The observe-orient-decide-act (OODA) model is utilized in the military to quickly respond in changing and unpredictable environments. Just as in the OODA model, with each cycle, take what you observe and learn and feed that learning into the next cycle for planning and implementation.

P3: Implement Data-Driven Decisions

Connection to Other Principles	P2: Apply Multiple Horizons of Planning
How the Principles Work Together	The backlog is defined at multiple levels of decomposition (epic, feature, user story, task) and is planned within multiple horizons of planning. As backlog items get closer to implementation, they are further refined. Each item contains a description, who needs the capability, business benefit, and acceptance criteria. As backlog items are demonstrated, the objective evidence oft what is and is not working is fed into the next planning cycle.

P6: Establish Cadence and Synchronization

Connection to Other Principles	P7: Integrate Early and Often
How the Principles Work Together	We bring the understanding of cadence and synchronization with us into the cyber-physical world as product teams plan, develop, and deliver system capabilities. Together they define the standard of repeatable planning sessions, large-system integration, and demonstrations of integrated working capabilities. With large cyber-physical solutions, these demonstrations are implemented in a hybrid manner with a mixture of digital and physical artifacts.

As your organization defines the cadence and synchronization points, communicate the need for CI at the system level as early as your environment can enable. Teams that integrate and demonstrate functionality together will want to be on the same cadence so their synchronization points align. |

Table 13.3: Principles Focused on Execution

P4: Architect for Change and Speed

Connection to Other Principles	P1: Organize for the Flow of Value
How the Principles Work Together	Intentional modular architecture with standardized interfaces reduces the number of dependencies, which enables the flow of value to stakeholders. The product backlog defines new business-facing functionality as well as enablers to evolve the architecture. Organize teams around the flow of value in concert with your architecture. One approach when considering our CubeSat example is having stream-aligned or complicated subsystem teams implementing capabilities such as attitude control, attitude determination, and attitude estimation and filtering while having a platform team build the infrastructure needed to run the software.

P5: Iterate, Manage Queues, Create Flow

Connection to Other Principles	P1: Organize for the Flow of Value
How the Principles Work Together	Teams are organized around value streams to improve the flow and delivery of value. This yields the benefits of iterative development and incremental delivery. While iterative development is often thought to be for software development only, this is not the case, as we have conveyed throughout this book. Embedded systems and hardware teams are now engaged in iterative development cycles. Players in the space industry have demonstrated this advantage. For example, Rocket Lab has "demonstrated the ability to support rapid integration and short notice customer-driven changes in launch schedule, inclination, and launch site."[9] As you are getting started, organize your teams around the value stream, create product backlogs that align with your cross-functional Agile team structure, and set up the tool environment that enables iterative development across software and hardware with feedback loops in manufacturing and with the customer.

P7: Integrate Early and Often	
Connection to Other Principles	P3: Implement Data-Driven Decisions P5: Iterate, Manage Queues, Create Flow
How the Principles Work Together	The goal of integration is to ensure that all of the systems being developed can communicate and share data effectively. Integrating early and often is not only from a software perspective; it means integrating at the system level. Integrating early and often not only buys down risks while ensuring fit for purpose, but when coupled with iterative demonstrations, it provides real-time information and data about what is working or not working. Metrics are reviewed to understand the current system state. As we manage and improve flow, integrated tool sets and dashboards generate these metrics, which can be used to see where the bottlenecks are in the system. Build the system iteratively, integrate early and often, improve flow, and use data to understand the current state and to make decisions on the next steps to take in developing the solution or improving the flow.
P8: Shift Left	
Connection to Other Principles	P3: Implement Data-Driven Decisions P5: Iterate, Manage Queues, Create Flow
How the Principles Work Together	Through experience, we have found that reviewing requirements through a lens of how they will be validated and verified has shown to ensure executable requirements. As you get started with a shift-left implementation, define your test strategy to incorporate a shift-left testing mindset. Acceptance tests are written before development begins. Shift-left manufacturing creates regular feedback to optimize test processes by identifying areas that are time-consuming or difficult to test. Verification of designs happens regularly. Through experience, we have found that reviewing requirements through a lens of how they will be validated and verified has shown to ensure executable requirements. As you get started with a shift-left implementation, define your test strategy that incorporates a shift-left testing mindset. Shift-left manufacturing creates regular feedback to optimize test processes by identifying areas that are time-consuming or difficult to test. Verification of designs happens regularly.

Table 13.4: Principles Focused on Continuous Improvement

P9: Apply a Growth Mindset	
Connection to Other Principles	All principles
How the Principles Work Together	A growth mindset requires continuous learning and relentless improvement. Organizations should continuously measure flow and reduce bottlenecks. Engage the customer regularly and gather feedback to relentlessly improve the product and process. Employ retrospectives at all levels and have a defined process for prioritizing new backlog items created from the retrospectives. When there is a systemic problem, take action to determine root causes and the next opportunities. These improvements should be made visible to the organization through dashboards and events. Never stop learning. Embrace change.

As you continue your Industrial DevOps journey, you will continue to experience how these principles work together to enable your teams to build better systems faster. Implementing only one or two principles, while it may address a need in your organization, will provide limited benefits.

GETTING STARTED

- Establish a leadership focal point.
- Define "as is" and "to be."
- Hold leaders accountable for providing line of sight to business objectives.
- Leverage change management best practices.
- Utilize the Industrial DevOps principles to build out a road map.
- Manage your transition iteratively and incrementally, reviewing and evolving your road map continuously.

QUESTIONS FOR YOUR TEAM TO ANSWER

- Has your organization captured the value stream for the product? Before you start iterating on designs and development, be sure the teams are organized around the value stream.

- Review the Industrial DevOps framework and the six steps. In which areas is your organization the strongest?
- Read through Westrum's Organizational Typology Model. Which category best describes your organization? What evidence leads you to this conclusion?
- Under Step 3: Build the Foundation, what has your organization already done to build a foundation? What would be the next opportunity?
- Step 4: Organize provides many opportunities for improving how work and people are organized to develop and deliver customer value. In which areas is your organization doing well? For areas to improve, is there low-hanging fruit that you can take advantage of in the near term?
- What might your desired future state look like?

COACHING TIPS

- Ensure you create a leadership focal point for the change who has been educated in Industrial DevOps principles.
- Ensure business objectives are visible to your entire organization.
- Do not underestimate the difficulty of changing culture; be intentional about architecting new culture.
- Take a day or two to conduct a workshop with your leadership team to identify the current strengths and weaknesses, as well as opportunities and threats. Define your current state and the desired future state. Based on the collective input, prioritize the next steps.
- Build a road map to the future state.
- The road map needs to be continuously evolved based on the changing needs of the organization.

CHAPTER 14

BARRIERS TO CHANGE AND ADOPTION

The barriers to change and adopting new ways of working are many. It is difficult and sometimes feels impossible to change. Take the experiment of "The Backwards Bicycle: What It Takes to Unlearn Old and Learn New Habits."[1] A single change in the working mechanics of a bicycle caused one man to take eight months to learn how to ride the new bicycle. It was a significant amount of time to unlearn one practice (the way the bike is supposed to operate) and learn a new practice (the new process for operating the bike). And what is even more interesting about the story is that once he had mastered the "backwards bicycle," he then found it equally hard to go back to riding a bicycle the regular way as initially designed.

With determination and practice, a new skill can be learned. What makes it so difficult? It isn't just motivation. There is also a neurological reason it takes so long for us to unlearn and gain new learning and new habits. Learning reconstructs and builds the neural network of the brain. This reconstruction, as one learns new behaviors and skills, takes time, consistency, and practice.[2] The more an individual practices a routine, the stronger the neuron connection and the stronger the habit becomes. Embracing new ways of working is as much physical as it is motivational.

Test this out. Change one of your practices and see how difficult it is. Move something you reach for every day. Or try using your non dominant hand for simple tasks for a day. I guarantee it will take a while for your brain to learn the new routine. I (Suzette) had a similar experience with my family. In my home office, I had moved the power strip for my laptops to be plugged into the outlet that was also controlled by the light switch next to the door of the room. You turn off the switch, you turn off the electricity powering my laptop. My family could not seem to learn to stop turning off the switch when they walked out of the room. They were demonstrating the power of habit. Eventually, I took tape and put it over the switch.

Now they were not able to flip the switch even if they tried. They were willing to follow the process. However, their old habits kept interfering. We even laughed about it at times because they really were trying to follow my request but they struggled to establish a new habit. To help them, I had to change the environment. Now, I don't have the tape on the switch anymore. But, yes, it did take months to learn, and, yes, those old habits still pop up now and then and I find my laptop without power. This was a simple example. Think how much more difficult it is when it comes to learning new ways of working. Learning and applying new skills is one of several barriers organizations encounter when taking on and scaling any change initiative.

The following sections address multiple barriers based on research, our more than forty years of combined experiences with change in large organizations, and contributions from our industry colleagues. Overcoming these barriers will require new learning, practices, and time.

Table 14.1: Barriers and Challenges to Industrial DevOps Adoption

List of Barriers and Challenges to Industrial DevOps Adoption
Lack of consistent implementation of Industrial DevOps–related processes and practices.
Lack of a growth mindset and hesitancy to rethink current ways of working.
Lack of skills and experience hinders the scaling of Industrial DevOps.
Lack of engagement and organizational alignment leaves people unmoved.
Challenges with complexity and dependencies and scaling across teams of Agile teams.
Challenges with regulated environments.

These barriers are encountered in a variety of domains and across different-sized organizations. As you scale to cyber-physical systems, these challenges become magnified and more difficult to overcome because of system complexity and more people involved. Be intentional in your adoption and continuous improvements of Industrial DevOps principles. Have a strategy and invest in the strategy. Be prepared to inspect and adapt to the changing environment.

BARRIER: Lack of Consistent Implementation of Industrial DevOps–Related Processes and Practices

The number of people involved in building and delivering cyber-physical systems is extensive. It often includes hundreds to thousands of individuals working toward a common product vision and understanding. In 2021, the *15th Annual State of Agile Report* reported that 46% of companies indicated that inconsistent processes and practices are a barrier to adoption within their company.[3] This lack of consistent implementation of foundational practices provides inconsistent results. This isn't meant to imply that everything has to be exactly the same; context matters. When building large, complex cyber-physical systems, there is a need to understand both principles and how different practices meet the intent of those principles. Once we identify what works within the local context, it is critical that we standardize the foundations with a level of consistency that brings together alignment of common language, shared mental models, and practices that enable successful systems level integration.

The 2022 US Government Accountability Office (GAO) *Weapons Systems Annual Assessment* interviewed several Department of Defense (DoD) programs, some of which were also implementing Agile, DevOps, or iterative approaches. They discovered many of the cases were not fully implementing recommended practices, such as early and continuous delivery of software to users. As a result, these programs were experiencing much longer lead times to delivery.[4] Without consistency, the organization cannot gather metrics to understand the state of the product, nor can they visualize the bottlenecks in the systems, making continuous improvement activities futile. A lack of consistent implementation of Industrial DevOps practices impedes progress, creates misunderstandings and communication barriers, and results in significant delays in the delivery of value to the end users.

To overcome this barrier, define your organization's operating model and the shared practices across teams building the integrated system. These teams benefit from having a standard in place that creates a shared understanding as to how they coordinate across the multiple horizons of planning, how they align around cadence and synchronization points, and how they use standard tool environments. It is not enough to have an integrated digital environment. Teams need to know how the environment is to be used and the standards of that environment. Also, having clearly defined roles and responsibilities creates clear ownership of how decisions are made. Start by selecting a Lean-Agile framework that scales to the needs of your organization. Then customize it to meet the needs of your organization while making it as simple as possible. With the attrition concerns many companies are experiencing, building from an industry standard is essential as it reduces the learning curve for new hires and enables them to produce value more quickly.

BARRIER: Lack of a Growth Mindset and Hesitancy to Rethink Current Ways of Working

Companies, like individuals, are subject to the principles of *prospect theory* in making decisions. Prospect theory (see Figure 14.1) was initially developed to explain individual decision biases; however, its insights also apply to organizations because, after all, organizations are made up of people. One principle of prospect theory is something known as loss aversion. The principle states that individuals are more sensitive to losses than gains. The negative utility associated with losses is believed to be more significant than the positive utility derived from equivalent gains. Coupled with this finding is that when things are going well for people, they are typically more risk averse.[5] Change will be viewed as a risk to their past and current success. As a result, when a shift is happening around them, they are still likely to hold to those past experiences and beliefs and become complacent.

As a result, companies that have been successful for decades or more typically have a much harder time changing their ways of working than new companies on the scene. It's natural to want to continue using what has been successful in the past. It can be hard to let go, and feelings of uncertainty begin to emerge. Unfortunately, in this new digital era, things are advancing at a faster pace than we've ever experienced. As we build more digital capabilities and automate processes, faster change happens. Leading in the digital age requires a growth mindset. Leaders learn. They ask questions. They don't claim to have all the answers but surround themselves with talented people who help formulate the answers. They have the ability to pivot when the environment around them calls for a different approach: "You cannot be the same, think the same, and act the same if you hope to be successful in a world that does not remain the same."[6]

Past successes can lead to complacency and impact the desire to change. While Agile practices have been around since the 1990s, organizations are still struggling when it comes to wider adoption. A survey conducted in 2021 revealed that 46% of responders stated that general organization resistance was a key barrier to wider adoption.[7] Similar to Dikert's research, it was discovered 38% of organizations implementing Agile at scale found that general resistance and management being unwilling to change was a key barrier.[8] If you are experiencing resistance, take time to understand why. The resistance could be lack of clarity around expectations and understanding what specifically they need to do. A similar idea was stated by W. Edwards Deming when it comes to quality management: "It is not enough that top management commit themselves for life to quality and productivity. They must know what it is that they are committed to—that is, what they must do."[9]

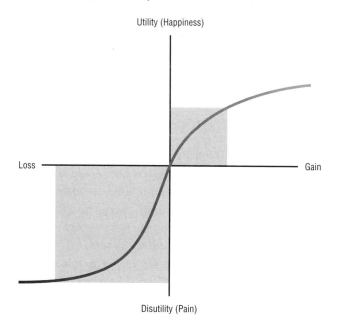

Figure 14.1: Prospect Theory
Source: Charlotte Nickerson, "Prospect Theory in Psychology."

Leading in the digital age requires a growth mindset. With changing priorities and new technologies and tools on the rise, we need to apply an attitude of continuous, lifelong learning. As a leader, continue to broaden your perspectives and continuously learn. Lead the way by demonstrating to your team learning in action. Since you are already reading this book, you are demonstrating a growth mindset. Congratulations! This is a great step. Encourage your teams to adopt a growth mindset as well. The next time you are in a meeting with your team, ask what they have learned or read lately that they would like to share with the group and how that learning might be applied.

With digital capabilities on the rise, continue to upskill talent. Support talent and skill development through training, communities of practice, book clubs, encouraging local meetups, and on-the-job coaching. Support comes in many forms. It can range from something as simple as sharing information to communicating, creating, and investing in learning opportunities.

Remember, under stress people typically revert to their earlier learned practices.[10] Coaching and mentoring reduce this risk. Executive management has to assure the provision of comprehensive and functional training, mentoring, coaching, and learning at all levels of the organization. Industrial DevOps as a mindset is much more important than the management methodology itself and leads to successful transformation.

Leaders and team members can help each other apply a growth mindset by asking questions that cause us to "think again." Some tips given by Adam Grant, author of *Think Again*, include:[11]

- **Ask *how* rather than *why*.** By asking *how*, we might move forward with an idea that otherwise might not be expressed and can help identify where there are gaps or limited understanding in views being discussed. Depending on how the question is stated, *why* conversations can lead to defensiveness versus the exchange of new ideas. Framing the question is important. For example, instead of asking "Why should we do this?" a simple reframing to "What are the anticipated benefits?" will likely lend itself to a more productive conversation.
- **Practice persuasive listening.** This means listening more than talking and asking more questions than offering statements. It helps team members formulate and more clearly articulate their views.

As leaders, let's foster a deeply rooted personal commitment to continual growth and development, both in ourselves and within each member of our team. Encourage lifelong learning and support others on that journey. Not only will fostering a growth mindset lead to greater personal, team, and organizational outcomes, it will also help everyone achieve their fullest potential.

BARRIER: Lack of Skills and Experience Hinders Scaling Industrial DevOps

The adoption of Industrial DevOps principles requires integrated tools, new technologies, and modern ways of working. To successfully execute and deliver value, teams need to be trained on the new skills required in this new environment. It is not uncommon for individuals who have not been trained to think being "Lean" or "Agile" equates only to the use of tools such as the Atlassian tool suite or only having daily stand-ups. We have experienced this misunderstanding too many times. The *15th Annual State of Agile* survey reported that 42% of responders identified lack of skills and experience as a key barrier.[12] In Dikert's research on barriers at scale, 31% of the results indicate a general lack of investment in coaching and training with ever-growing workloads.[13] And the 2022 GAO *Weapons System Annual Assessment* concluded one of the challenges in delivering capabilities faster is the lack of training in Agile for DoD acquisition staff and lack of existing enterprise infrastructure to support implementation.[14]

We have had many experiences where leaders have wanted to adopt Lean and Agile ways of working but have skipped educating themselves and their teams. After all, they are smart people; they should be able to figure this out. While they are smart people, a radical shift in how someone has been working for many years is not easy. We've seen teams read an article or book on Agile, and then change was just expected to happen. A fundamental change in how we work requires learning new skills, gaining experience through the application of the learning, and providing support via an experienced coach.

In your environments, you are probably witnessing something similar. The question is how to upskill talent and provide continuous learning and growth opportunities. Training is not a one-and-done activity. Even with successful training, the challenge is often in the implementation. The learning of new behaviors to replace old habits takes time. Reinforcement is necessary.

While lack of skills and experience continues to be a concern, research is identifying success patterns. We have modified "Agile" in the statements below to "Industrial DevOps" as we believe the same success patterns are required for the building of cyber-physical systems. These success patterns include:

- Management support and training management on ~~Agile~~ Industrial DevOps practices.
- Providing training on ~~Agile~~ *Industrial DevOps* methods and coaching as teams learn.
- Effective training and workshops and ~~Agile~~ *Industrial DevOps* coaching support.
- Having change agents and champions actively engaged in the transition.
- Early adopter pilot programs and communities of practice to build expertise and learning.
- Coaching for organizations transitioning from waterfall to ~~Agile~~ *Industrial DevOps*. Coaching can cover multiple areas, such as Agile and Lean practices, CI/CD pipeline, model-based engineering, supply-chain integration into the process, or wherever your team is embracing a technology and requires additional on-the-job support.

BARRIER: Lack of Engagement and Organizational Alignment Leaves People Unmoved

As a result of the size and complexity of cyber-physical systems, there are likely hundreds of people involved in the solutions; in some instances, for things like planes and large spacecraft vehicles, there could be thousands involved, covering all areas of the value stream. Therefore, when the organization decides they want to be more responsive to change and they want the ability to quickly embrace new digital capabilities, it is important for leadership to clearly articulate why change is happening and how they will provide the needed support so people have what they need to be successful. Leaders must communicate, align, and engage the workforce in this process. Research studies have confirmed that lack of engagement and organizational alignment serve as a major barrier to the adoption of change:

- 41% of responders to the *15th Annual State of Agile* survey suggest absence of leadership participation in the effort and inadequate management support and sponsorship as significant barriers to Agile adoption.[15]
- Functions are unwilling to change to a new way of working and experience challenges in adjusting to the incremental delivery of value cadence.[16]

- According to the GAO 2012 report, Agile guidance was not clear, and therefore we are getting suboptimal results. In addition, the federal reporting practices do not align with Agile. This provides an opportunity for the organization to improve their engagement levels and strengthen program delivery.[17]

Transforming ways of working with large groups of people requires significant coordination and leadership. Leadership clearly articulates the vision and the path forward. Use a change management framework to help shape your transformation journey. There are several change management frameworks, such as Prosci's ADKAR model and John Kotter's change model. John Kotter's Eight Steps for Change model provides an effective foundation for communicating why change is happening and the vision for a path forward.

Other success patterns include the following:

- Leadership engagement. Have a spokesperson/s for the change effort and a roll-out team to support line and project managers.[18]
- Build a powerful guiding coalition for change. Create shared commitment with leaders that have the authority and ability to change the existing system: "In successful transformations, the chairman or president or division general manager, plus another 5 or 15 or 50 people, come together and develop a shared commitment to excellent performance through renewal."[19]
- Many organizations set up a coaching homeroom and Lean-Agile Center of Excellence that provide guidance and support to new teams.[20]
- Positive factors include open communication, employee involvement, and empowerment.[21]

The Prosci ADKAR model provides another perspective for organizational change that addresses the needs of specific user groups being impacted by the change. It outlines five areas that need to be addressed for the user groups in order for the change to happen. The five areas the organization must plan for are building Awareness, Desire, Knowledge, Ability, and Reinforcement (ADKAR).

The integration of the Prosci ADKAR model and Kotter's model will help your organization lay out a change management plan and implementation road map that addresses a variety of needs specific to your organization, putting you on a path to success.

BARRIER: Challenges with Complexity, Dependencies, and Scaling across Teams of Agile Teams

Cyber-physical systems are complex, adaptive systems. Building these large systems requires all parts of the organization to be involved in the development and production of the system. Beginning with market understanding and strategy shaping across the functions to include finance, law and contracting, supply chain, business areas, engineering, manufacturing, human resources, security and compliance, and anyone else directly involved in ensuring a successful product delivery, this level of scaling magnifies the need for dependency and complexity management.

Complexity and dependency management impact the speed and teams' ability to deliver value. When organizations continue to align the work around functional areas, this creates more dependencies, hand-offs, and bottlenecks. Large organizations must be proactive to integrate their non development functions into product value streams. The traditional practice of organizing around functional areas is built from a mindset that results in functional silos, which are still alive and well in many organizations. The result is long lead cycles created by dependencies and hand-offs between functions, communication barriers, and increased risks, especially at systems integration, which is one of the greatest bottlenecks in the system.

There are also challenges with supply chain and procurement practices, which might not be updated to support rapid response to change and shorter delivery cycles. Engaging with suppliers to innovate and deliver faster can be a procurement challenge if it is not addressed in their contracts. In addition, engaging with *many* suppliers creates a level of complexity when it comes to coordinating frequent or early integration and test opportunities and wanting to streamline processes for more efficiency. At the same time, it is important to embrace the expertise suppliers bring to large systems development. They have special technologies or provide critical parts required for complete production of the system. Understanding how to streamline and effectively engage with your suppliers is part of your Industrial DevOps strategy.

Dependency management is critical to the success of your Industrial DevOps implementation and is best addressed with two of the principles we discussed in this book, which are organizing around value and architecting for speed:

We placed a focus on the business and transforming how we work first and foremost. We used that as a catalyst to begin to drive agility through the organization—speed of decision making, clarity of focus on purpose, showing that we can deliver incremental value—without forgoing the larger picture.[22]

The Industrial DevOps principle "Organize for the Flow of Value" ensures we put a cross-functional team together capable of completing all of the work across the value stream.

The Industrial DevOps principle "Architect for Change and Speed" helps reduce dependencies through modularity of software and hardware system components. Have well-defined interfaces and integration points. Standard interfaces allow more independent design evolution.

The US government continues to make progress in modernizing its procurement process. In April 2022, a framework under the AGILE Procurement Act was created to improve the procurement process. Under this framework:

> . . . the administrator for Federal Procurement Policy in the Office of Management and Budget will offer guidance on 'the availability of streamlined and alternative procurement methods' for IT and communications technology, such as simplifying procedures. . . . the act offers a plan to address red-tape and barriers to entry for federal contracting, such as examining past performance to help expand the pool of eligible contractors. In doing so, the bill seeks to reduce barriers to entry for entities that are trying to work with the federal government.[23]

Continuously advance the technologies and processes you use to build complex systems and innovate faster:

> [Cyber-physical systems] are becoming data-rich, enabling new and higher degrees of automation and autonomy. Traditional ideas in [cyber-physical system] research are being challenged by new concepts emerging from artificial intelligence and machine learning. The integration of artificial intelligence with [cyber-physical systems] especially for real-time operation creates new research opportunities with major societal implications . . .[24]

BARRIER: Challenges with Regulated Environments

Most cyber-physical systems are highly regulated. For example, space vehicles are subject to regulations and standards such as those defined by NASA or the European Cooperation for Space Standardization (ECSS) safety standards. Therefore, before they can be certified for use, they have to meet a number of regulations, which often results in much longer lead times for product delivery.

Many people have the misperception that Agile and DevOps do not lend themselves to regulated environments, but nothing could be further from the truth. We build in the constraint at the start of delivery as opposed to bolting on after the product has been built. This approach allows us to use all of the tools at our disposal—such as behavior-driven development, automation, continuous integration, and rapid feedback loops—to ensure we build safety and security in from the start. Just as we have learned from Lean about the importance of building in quality, we build in safety.

When you first begin, you may experience resistance from the regulating organizations because they are used to waiting until everything is complete before providing feedback. This is an area where the industry is continuing to evolve.

CASE STUDY: The US Department of Defense Makes a Case for Agile Acquisition[25]

The United States Department of Defense (DoD) is one of the most highly regulated environments in the world. Year after year, the Government Accountability Office performs an assessment of the largest weapon programs. The US has an aging infrastructure with a need to deliver capabilities to the war fighter at a much faster pace. The industrial base has frequently stated that the waterfall acquisition process is a barrier to delivering needed capabilities faster.

The DoD decided to pilot a new Agile acquisition approach in an effort to deliver software capabilities faster. The pilot included seven programs that represented a cross-section of the military departments, as well as a mix of weapons and business systems. The pilot lasted one year. In October 2019, the DoD reported on the completion of the pilot:

- All participating programs had successfully adopted Agile or iterative software practices.
- Participating programs delivered working software far faster than similar traditional acquisition programs.

Given the results of the pilots, the DoD issued an updated Agile software acquisition guidebook in February 2020 that included best practices and lessons learned from the pilot. Example best practices included the need for:

- Agile coaching for programs transitioning from waterfall to Agile, which was deemed critical.
- Sustainment planning to address activities such as designing for modularity and managing technical debt.

GETTING STARTED

- Recognize the barriers to adoption and address them in your transformation.
- Communicate the reason for the change and engage others across the organization in the process.
- Leadership engagement and the mentoring of others are paramount for success. If your organization is brand new to Industrial DevOps, find a pathfinder or early adopters who are willing to embrace new ways of working. Start small and grow.
- Build in quality and safety standards.

QUESTIONS FOR YOUR TEAM TO ANSWER

- What are the top barriers your organization needs to address?
- What change management model will you use? Will you use only one model? How will you use the change management model?
- How will you deal with resistance in your organization?
- How are you demonstrating a growth mindset?
- What are the expectations of leadership?
- What is your training approach to learning new tools and processes?
- How will you support new teams during the transition to Industrial DevOps?

- What steps can you take to build in quality and safety?
- Have you defined and communicated the expected standard practices? How have you allowed for team autonomy?

COACHING TIPS

- Along your Industrial DevOps journey, you will encounter many barriers. We have only listed the most common that we have experienced. Whenever you encounter a barrier, apply a growth mindset and start by thinking how the barrier might be overcome.
- As a leader, you must commit to continuous learning. How are you learning to become a better leader? How often do you visit with the people who do the work to understand how things are working? Do you talk more or listen more? Encourage lifelong learning and support others on that journey.
- Take time to reflect on what you have learned recently. Become self-aware. This can help you lead yourself and others more effectively.
- Define the standards and train people on how they are going to be used. For example, it is not enough to train people on how to model; they must also understand how the models will be used to support your environment. Have a defined training approach that offers learning opportunities in a variety of ways to address different learning styles.
- Industrial DevOps invites new ways of working that involves people, processes, and tools. Remember respect for people and culture comes first.

Conclusion

> Change is a common thread that runs through all businesses regardless of
> size, industry and age. Our world is changing fast and, as such, organiza-
> tions must change quickly too . . .
>
> —Kurt Lewin, Change Management

Our path to Industrial DevOps started twenty years ago. Over the years,
we have worked in a variety of settings for different customers across a
myriad of products and integrated systems. We have experienced successes.
We have struggled. We have continued to learn along the way. Based on
this journey, we have collected and shared with you success patterns that
define the Industrial DevOps principles. As you embark on your Industrial
DevOps journey, there are three critical insights to guide your journey:

1. Industrial DevOps applies to the entire system.
2. Digital capabilities enable fast feedback loops and shift-left
   practices.
3. People and culture are the key to success.

Industrial DevOps Applies to the Entire System

Industrial DevOps applies to the entire system (the organization and the
cyber-physical system). Thus, it is essential to apply systems thinking and
look holistically across the system for improvements and opportunities.
This has been a primary lesson for us and continues to be validated as we
exchange with companies from around the world who build cyber-physical
systems. While Lean, Agile, and DevOps have existed for decades, it is the
weaving together of these practices across the value stream that improves
the flow and delivery of value. With the rapid advancement of digital

capabilities and tooling, now is the opportune time to embrace Industrial DevOps.

As you begin your Industrial DevOps journey, remember to start with "why." Why are you making this shift? What is the imperative for your organization that makes now the right time? According to Simon Sinek in *The Infinite Game*, for organizations to stay in the game, it is no longer about "who wins or who's the best" but rather about building organizations that are healthy and able to survive for decades to come.[1] Work across the organization to define the business outcomes you are after, define your future state, and create your road map for getting there. You will need to adjust along the way—not because your initial plan was wrong but because the ecosystem around you continuously evolves. Industrial DevOps principles enable you to inspect, adapt, and course correct so you can take advantage of emerging markets, new technologies, and changing priorities of the business. Take with you the coaching tips that we have provided with each chapter and use them as a guide to take your next steps. Embrace a growth mindset. Commit to continuous learning.

Digital Capabilities Enable Fast Feedback Loops and Shift-Left Practices

The advancement of digital capabilities and technologies has created an environment in which we can now apply what we have learned from the decades of software development to the world of hardware and manufacturing. Digital capabilities have enabled shift-left practices in which testing, compliance, and design for manufacturing are now part of the integrated and iterative development process. This shift results in higher-quality products and reduces rework downstream by building in quality and inspections along the way through integrated system demonstrations. Teams' processes enable short feedback cycles in which they get feedback on cadence from other teams and stakeholders such as customers, business leads, and users and can quickly adjust based on the feedback received. Status is not focused so much on task completion, as now it focuses on demonstrations based on defined acceptance criteria.

With physical hardware, we are able to shift left as we move physical development and testing into digital environments. The development of physical systems continues to take advantage of emulators, simulators, and prototypes along with the advancement of emerging capabilities such as 3D printing, additive manufacturing, digital threads, digital twins, and virtual reality and augmented reality (VR/AR). Digital capabilities impact not only how we develop cyber-physical systems but also the factory used to build the system. The emergence of Industry 5.0 and the smart factory brings in a wide array of digital capabilities. New capabilities are tested early, and with improved safety, through the use of automation, robotics and autonomous systems, VR/AR, AI, and more. The range of possibilities to improve operational efficiencies and deliver faster continues to emerge. Shifting left enables early validation of designs, reduces risks, and builds in quality to build systems better and faster.

People and Culture Are the Key to Success

Who builds these better systems faster? It is the people. Therefore, the people and culture of the organization are the key to success. Meet people where they are in their Industrial DevOps journey and take time to understand their existing culture. In many instances, you will discover that it is not that people aren't willing to adopt new ways of working. The hesitancy is more often because they don't understand the reason for the change, they don't know what it is they are being asked to do, there is fear of failure, and they are not given the time and support they need to learn. By creating a culture of continuous learning, you can begin to address these concerns. Listen to what they need and provide them with the resources, training, digital tools, and environment necessary to do their job.

With the complexity of large cyber-physical systems, it is recommended that adoption of Industrial DevOps principles starts with one of the smaller value streams of the system and then grows. Build off the smaller successes to generate more success. Our experiences align with Jonathan Smart's thinking as described in *Sooner Safer Happier*:

> . . . people have a limited velocity to unlearn and relearn. You cannot force the pace of change, even if you think that you are. The outcome will be new labels on existing behaviors, the robotic maneuvers of Agile, people in an agentic state waiting for the next order, and little actual agility. . . . *apply an agile approach to agility and achieve big through small.*[2]

Where to Now?

Digital capabilities and new technologies are growing at an unprecedented rate. As we write this book, we continue to learn more every day about advancements around us. The world of artificial intelligence (AI) has picked up speed, with the impact still largely unknown. Based on current observations, it will change how we build systems, from development through production and delivery. It will change how we learn. Rote tasks will become more automated. Higher-level thinking skills and creativity of our teams will become increasingly important. For instance, a developer may use AI to help design test cases or to find errors, and AI will learn to ensure those errors are not repeated.

However, the developer needs to know what questions to ask and what processes need to be followed and to be able to validate the accuracy of the information received. And this is just a small example of the changes swarming upon us. Machine learning, AI, autonomous systems, robotics, continued enhancement of VR/AR, and more are evolving each day. Continuous learning and adapting to changing environments is required for survival and will provide opportunities that we have yet to explore.

We know this journey is not finished, because the world has not stopped evolving. The "digital age" has only just begun.

The Next Step Is Yours

We have shared and demonstrated the Industrial DevOps principles throughout this book and have described various tools and techniques you can use to help you. We have provided examples of companies demonstrating these principles, and you can see the world around you changing. Reflecting on what you have read, define your next steps in adopting Industrial DevOps principles to build better systems faster.

As we said at the beginning of this book, "Companies that solve this problem first will increase transparency, reduce cycle time, increase value for money, and innovate faster. **They will build better systems faster and become the ultimate economic and value delivery winners in the marketplace.**"

The next step is yours.

Thank you for being part of our learning journey.

APPENDIX A

CUBESAT 101

In order to build a common mental model while discussing the Industrial DevOps principles in this book, we frequently use the example of a CubeSat (a cube-shaped miniature satellite). We use this example because they are cyber-physical systems that are relatively simple to understand. If you're not familiar with CubeSats, this section will break the basics down for you.

Machine-based satellites are used for a wide variety of purposes to do everything from weather forecasting to using a global positioning system (GPS). Many of us make use of applications such as Google Maps in our everyday lives to obtain directions to go from one place to another.

Traditional satellite development has a high barrier to entry due to the expense of design and launch. According to GlobalCom, a typical weather satellite costs about $290 million to build and between $50 and $400 million to launch into orbit.[1] In the late 1990s, two professors from California Polytechnic State and Stanford wanted to help their students gain hands-on experience with engineering satellites. They introduced what is now referred to as CubeSats, which are miniature satellites that are also cost-effective. A typical CubeSat is a ten-centimeter, or four-inch, cube with a mass of less than 1.5 kilograms (three pounds).

CubeSats are small but mighty cyber-physical systems (see Figure A.1), which is why we selected them for our example. They can be as simple or complex as we want to make them.

Interestingly enough, the use of CubeSats has exploded over the last fifteen years, with a target demand expected to reach $857.39 million by 2030.[2] They have primarily been flown in low Earth orbit (LEO), but now CubeSats are moving out to deep space. You can build a CubeSat for $1,000 with a launch price of between $10,000 and $40,000. This has changed the game. Because of their size, they are exponentially cheaper to launch. And their defined technical standards make it easier for new players to enter the market. They can be grouped together to create larger capabilities, such as Starlink, a satellite internet constellation operated by SpaceX.

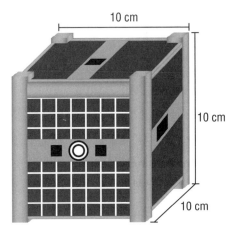

Figure A.1: CubeSat Dimensions

CubeSat Architecture

We are going to share a high-level architecture and design of CubeSat to make our examples throughout the book easier to follow. Our CubeSat components are illustrated in Figure A.2.

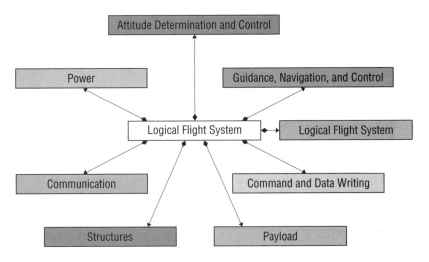

Figure A.2: CubeSat Logical Component Architecture

The description of each of the components is outlined in Table A.1.

Table A.1: CubeSat Logical Component Description

	Component	Description
1	Attitude Determination and Control	Detect and control orientation of CubeSat.
2	Guidance, Navigation, and Control	Navigation, which tracks current location; guidance, which uses navigation data and target information to determine where to go; and control, which accepts guidance commands.
3	Thermal Determination and Control	Sensors to detect and control temperature of CubeSat.
4	Command and Data Handling	Accept commands from ground and dispatch to the CubeSat.
5	Payload	Collect mission-specific data (weather, location).
6	Structure	Controls for the physical system CubeSat.
7	Communication	Transmit data between CubeSat and ground station on CubeSat flight data and mission data.
8	Power	Collect, store, and regulate energy.

Just as with traditional satellites, CubeSats still have to meet stringent requirements, such as using components able to withstand space conditions. The hardware components for our CubeSat are outlined in Figure A.3.

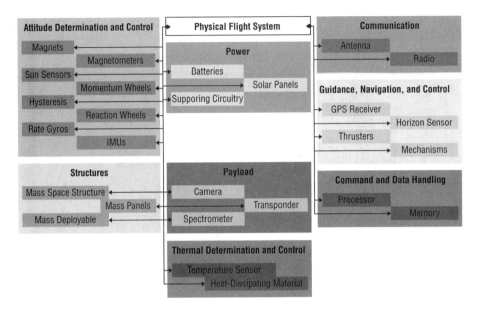

Figure A.3: CubeSat Hardware Component Architecture

In addition to the hardware architecture, we have outlined the software architecture for our simple CubeSat in Figure A.4.

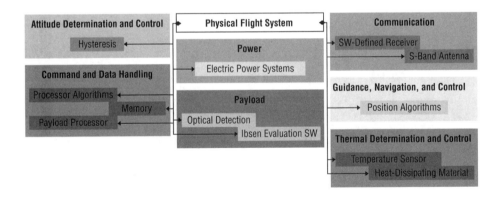

Figure A.4: CubeSat Software Component Architecture

The full physical system of interest is outlined in Table A.2.

Table A.2: CubeSat Physical Components

Subsystem	Component	Description	What It Does
Structure	Access Port	Physical surface inter-face to CubeSat when it is in the dispenser	Access your CubeSat with RBF pin
	Frame	Physical structure of CubeSat	Hold all of the CubeSat components
	Side Panel(s)	Physical structure of CubeSat	Provide access to compo-nents in CubeSat
Power and Thermal	Solar Sensor	Small, lightweight dig-ital sensor that detects UV and infrared light	Determines spacecraft body angles with respect to the sun
	Solar Panel	Physical aluminum panels that collect sunlight and convert into electrical energy	Recharge the batteries in the EPS
	Electron Canon	Used to oust excess electrons	Supports solar wind sail
	Electromagnet	Creates a magnetic field that controls the amount of electric current	Rotates spacecraft by controlling power
	Electrical Solar Wind Sail	Propellantless propul-sion system	Moves the CubeSat
	Electrical Power System (EPS)	Contains batteries that power the CubeSat	Powers the CubeSat
	Tether Wheel Motor and Electronics	A spool with long cables that are used for propulsion, momentum exchange, stabilization, or attitude control	Creates momentum to move spacecraft

Table A.2: CubeSat Physical Components Cont.

Subsystem	Component	Description	What It Does
	Tether Endmass	The endmass for the cables	Attaches to tether
ADCS	Command and Data-Handling System	Electronic circuit board that consists of an onboard computer with interface to all subsystems.	The brain for satellites, which handles all operations.
CDHS	Command and Data-Handling System	Electronic circuit board that consists of an onboard computer with interface to all subsystems.	The brain for satellites, which handles all operations.
Communication	Antenna	Copper adhesive tape	Transmit and receive signals for communication
	Radio	Electrical circuit board that transmits and receives UHF/VHF signals	Transmit and receive radio signals for communication
	Communication System	Electronics circuit board containing radio, transceiver, and beacon transmitter and radio	Communicates with controllers on ground or other CubeSats in space
Payload	Camera	Optical device to capture images	Obtains images

CubeSats are advancing space research across a variety of areas such as data transmission (internet accessibility), reduction of orbital debris, science and exploration, earth observation (such as predicting natural disasters), and more.

CubeSat Mission Scenario

CubeSats fulfill many missions. These might include space imagery to understand distances between various objects in space, space weather patterns, imagery and data collection such as tracking of endangered animals, weather events, physical environments, and a variety of science experiments by researchers.

The CubeSat mission we use in this book is to improve weather forecasting accuracy to improve the safety of life and property. In our scenario, we have a targeted customer who is interested in this data to improve their predictive models and provide better weather forecasting. The initial goal of the business is to iteratively launch two hundred CubeSats within the first fifteen months and reduce the lead time to twelve months. The CubeSats need to be replenished every twelve months while building enhanced capabilities. The initial launch has a limited set of capabilities, which will continue to be enhanced with each launch. We include enabling capabilities such as a digital twin instance for each satellite and artificial intelligence as part of the satellite network for advanced communications and data analysis. Due to the number of CubeSats in production and the advanced technologies and growth of the organization, we have eight small teams working on the attitude control systems, which includes the configuration, modeling, and digital twin capability. The business intends to use this mission as a starting point, with plans to extend the business over the next three years.

We selected this mission because it is understandable, regardless of one's background and experiences. The affordability of CubeSats is impacting the space industry, breaking down barriers to entry for small businesses, increasing the ability to innovate and experiment, and improving how we educate and grow the future workforce.

APPENDIX B

INDUSTRIAL DEVOPS BODIES OF KNOWLEDGE

Industrial DevOps is built upon several existing bodies of knowledge. These bodies of knowledge have been proven and validated for decades and are foundational to delivering cyber-physical systems. The bodies of knowledge include Lean, Lean Startup, Agile, DevOps, design thinking, systems engineering and model-based engineering, architecture, and systems thinking. A timeline of these bodies of knowledge is presented in Figure B.1.

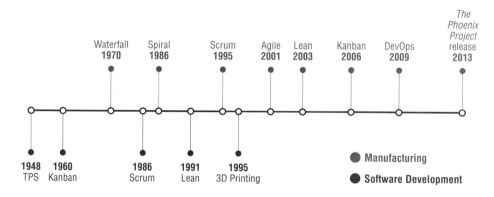

Figure B.1: History of Industrial DevOps Bodies of Knowledge

This appendix will introduce you to the multiple bodies of knowledge we pulled from in defining Industrial DevOps principles for cyber-physical systems. Please note that we are not advocating for one body of knowledge, framework, or method over the others. We suggest understanding all of the tools available in your toolbox and using the correct tools for the problem that you are trying to solve. In most cases, you will find a composite of multiple tools is the right answer (see Figure B.2).

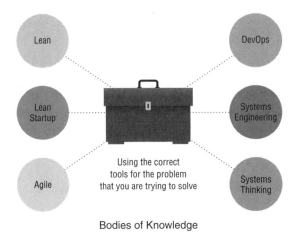

Bodies of Knowledge

Figure B.2: Industrial DevOps Bodies of Knowledge

Lean

According to the Lean Institute, Lean is a way of thinking about creating needed value with fewer resources and continuous experimentation.[1] Lean concepts date back to Venice in the 1450s, when a process was developed to sequence and standardize the process of galley shipbuilding for shipbuilders.[2] The Venetian Arsenal was said to be able to move ships through their entire production line in an hour.

In 1910, Henry Ford developed his manufacturing strategy for the automobile, where he arranged people, machines, equipment, tools, and products to obtain a continuous flow of production. While Ford was not the inventor of the automobile, what he did was invent an approach to improve the flow of work with the moving assembly line and the five-dollar workday.[3] The assembly line, the use of the conveyor belt, and streamlined practices improved the production of the automobile, making it more affordable for more people. Not only did his strategy result in more affordable cars, but the return to the company resulted in increased wages and improved living for the employees.[4]

The concept of Lean evolved with Taiichi Ohno in the 1950s based on the success patterns and principles of the well-known Toyota Production System (TPS). It was further enhanced by W. Edwards Deming's Total Quality Management System. In the 1990s, James P. Womack released *The Machine That Changed the World*, based on his extensive studies of the Toyota Production System in Japan.[5] Womack showed that the Toyota Production System was applicable to any company in any industry in any country. His team searched for a term that would describe its universal nature and the name "Lean" stuck.

Lean, illustrated in Figure B.3, has five key processes: identify value, map the value stream, create flow, establish pull, and seek perfection. The most important concept in the Lean process is customer value. Key benefits obtained from using the Lean process are increased efficiency, reduced waste, increased productivity, and increased product quality. The goals is to pursue perfection.

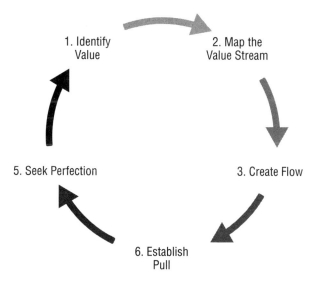

Figure B.3: Lean Production Cycle

Lean Startup

The concept of Lean Startup originated in the early 2000s with Steve Blank and Eric Ries and evolved into a methodology by 2010. Eric Ries went on to publish *The Lean Startup*, a book for entrepreneurs to use continuous innovation and learning to bring innovative products to market rapidly. Eric describes a build/measure/learn cycle, illustrated in Figure B.4, which is simply the scientific method reimagined for business.

Instead of one hypothesis, as we see in the scientific method, Ries promotes two hypotheses, which are a value hypothesis (What problem are we trying to solve?) and a growth hypothesis (How can I scale the benefit?). Next, we run an experiment with the fastest, cheapest way to validate the hypothesis. During the experiment, he recommends not asking for opinions but rather observing people's behavior about how they interact with your product. Based upon observations, adjust the product, with the goal being to complete the build-measure-learn loop as fast as possible. (The process he describes very much aligns with how we train algorithms in machine learning.)

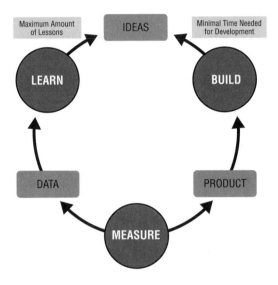

Figure B.4: Build, Measure, Learn

Another key concept that Ries defines is a minimum viable product (MVP), which he defines as a product with just enough features to learn from.[6] This concept is often misunderstood across many domains, where people interpret MVP as something releasable to the customer. For example, instead of a cardboard prototype or model that allows quick validation of our ability to fit various smartphones into a car compartment, we wait until we can actually release a physical product into the vehicle. This approach to the MVP delays our learning and reduces our ability to adapt.

Value Stream Management

The concepts of value stream management were published in the book *Project to Product* by Mik Kersten in 2018. *Forbes* defines value stream management as "a lean business practice that helps determine the value of software development and delivery efforts and resources."[7]

Value stream management leverages techniques such as value stream mapping that were popularized in the 1980s by James Womack and Dan Jones in regard to their work on the Toyota Production System in the 1980s.

Agile

Agile is not one single principle or practice but a product development life cycle, as waterfall is a product development life cycle:

- **Waterfall**: predictive, synchronous, phase-gate delivery mechanism
- **Agile**: empirical, iterative, incremental delivery mechanism

Agile is an evolution of an iterative and incremental approach to managing work that was first described in the 1930s, when the physicist and statistician Walter Shewhart of Bell Labs applied what he referred to as Plan-Do-Study-Act (PDSA) cycles to the improvement of products and processes.[8] Multiple practitioners, including W. Edwards Deming, further evolved the approach to developing products.

In 1986, Hirotaka Takeuchi and Ikujiro Nonaka published "The New New Product Development Game," where they compared product development to a game of rugby.[9] They discussed the team as a Scrum that operates as a single unit moving the ball down the field to accomplish their goal. They defined development as a holistic approach to building products that outline six characteristics: built-in instability, self-organizing project teams, overlapping development phases, "multi-learning," subtle control, and organizational transfer of learning. You can see the initial instantiation of this development life cycle had nothing to do with software.

As is common in all things, what is old became new once again when a group of software professionals collaborated to build a better approach to software development. This group defined the Agile Manifesto, illustrated in Figure B.5, to minimize challenges associated with software development.

Figure B.5: Agile Values

The Agile Manifesto was written in 2001 and promoted four core values: (1) individuals and interactions over processes and tools, (2) working software over comprehensive documentation, (3) customer collaboration over contract negotiation, and (4) responding to change over following a plan.[10] In addition to the manifesto, the authors agreed to twelve principles, outlined below in Table B.1, to back up the manifesto.

Contrasted with the waterfall phase-gate life cycle, Agile uses short, iterative cycles with frequent customer involvement to incrementally deliver products, resulting in increased adaptability, shorter schedules, reduced cost, increased transparency, and higher employee morale.

Table B.1: Agile Principles Defined

Twelve Principles		
Our highest priority is to satisfy the customer through early and continuous delivery of valuable software.	Welcome changing requirements, even late in development. Agile processes harness change for the customer's competitive advantage.	Deliver working software frequently, from a couple of weeks to a couple of months, with a preference to the shorter timescale.
Business people and developers must work together daily throughout the project.	Build projects around motivated individuals. Give them the environment and support they need, and trust them to get the job done.	The most efficient and effective method of conveying information to and within a development team is face-to-face conversation.
Working software is the primary measure of progress.	Agile processes promote sustainable development. The sponsors, developers, and users should be able to maintain a constant pace indefinitely.	Continuous attention to technical excellence and good design enhances agility.
Simplicity—the art of maximizing the amount of work not done—is essential.	The best architectures, requirements, and designs emerge from self-organizing teams.	At regular intervals, the team reflects on how to become more effective, then tunes and adjusts its behavior accordingly.

Since 2001, Agile has been known as a software development approach utilized to improve software delivery. Numerous benefits have been reported, including increased ability to adapt to change, reduced development schedules, reduced development costs, increased product quality, increased stakeholder transparency, and increased employee morale. Some of the most common frameworks include Scrum, kanban, Lean, eXtreme programming (XP), feature-driven development (FDD), dynamic systems development method (DSDM), and eXtreme manufacturing. Each framework provides a team structure, communication mechanisms, tools, and artifacts. The common characteristics of the frameworks are iterative, incremental, modular, time bound, simple, adaptive, transparent, collaborative, value focused, continuous feedback, and rapid learning.

While there are many common elements within the frameworks, there are many underlying practices, as illustrated below by the Agile Alliance's subway map.[11] Practitioners select the practices that support their solution needs. Many of the practices are interdependent on others. One practice enhances the effects of another. However, it is not recommended to implement them all at once or in their entirety.

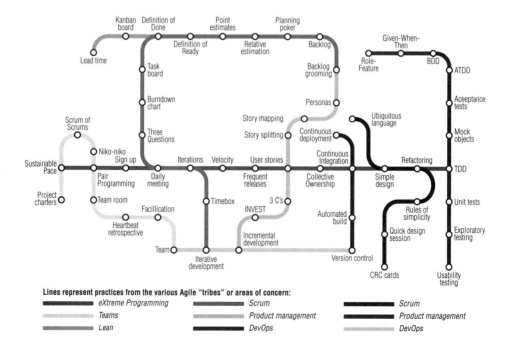

Figure B.6: The Agile Subway

Source: Carignan, Louis-Philippe, "Agile, Is It Just a Delivery Mechanism?"

The initial literature around Agile software development focused on small, cross-functional software teams that were colocated. The success that small Agile software teams achieved led to the question of whether Agile frameworks could be successfully scaled to support many interdependent teams. According to Digital.ai's *16th Annual State of Agile Report*, the Scaled Agile Framework (SAFe) is the most utilized scaling framework, with 53% of the scaling users reporting the use of SAFe. The second-most popular approach is Scrum@Scale, which increased in popularity in 2022 after years of decline. (The full list of popular frameworks is in Table B.2.)

Table B.2: Most Popular Agile Scaling Frameworks

Framework	Author	Date	Used
Scrum@Scale	Jeff Sutherland; Ken Schwaber	1996	9%
Large-Scale Scrum (LeSS)	Craig Larman; Bas Vodde	2005	3%
Agile Portfolio Management	Jochen Krebs	2008	3%
Scaled Agile Framework (SAFe)	Dean Leffingwell	2011	37%
Disciplined Agile (DA)	Scott Ambler	2012	3%
Spotify	Henrik Kniberg; Anders Ivarsson	2012	5%
Enterprise Scrum	Mike Beedle	2013	6%
Unknown/Other	N/A	N/A	23%

Similar to the team-level frameworks, each one of the frameworks provides organizational structure, communication mechanisms, tools, and artifacts. The common characteristics of the frameworks at scale—that is, requiring the coordination of many agile teams across multiple functional areas to deliver an integrated product—in addition to the team-level ones are team coordination, organization hierarchy, system architecture, dependency management, requirements management, value stream management, and the integration of non development functions.

Multiple papers and studies have compared and contrasted the scaling frameworks. Based on that analysis, we have found that at the practice level, there is a significant amount of overlap and agreement in the recommended practices. Therefore, the Industrial DevOps principles defined in this book align with any of the scaled frameworks.

DevOps

DevOps evolved from a series of events and was built upon movements that had come before, including Lean and Agile.

In 2009, Patrick Debois popularized the term *DevOps* at a Velocity event in Belgium. Development and Operations teams had a dysfunctional relationship, which resulted in large delays and defects. Agile software development exacerbated this existing problem. The two domains had never been aligned; Development teams are incentivized to deliver frequently, which maximizes change, and Operations are incentivized to keep the operation baseline stable, which minimizes change. Once software started being deployed faster, the relationship became untenable.

The solution itself was a simple one. Align Development and Operations with a common set of incentives, which was to deliver fast while keeping the operational baseline stable. The result of this simple change was shorter lead times, lower costs, and increased levels of quality. Now, over a decade later, there have been countless books to describe this cooperation between Development and Operations to deliver capability rapidly to the user, with the most impactful being *The Phoenix Project* in 2013.

For purposes of this discussion, we define DevOps as a "mixture of people, process, and technologies that provides a delivery pipeline enabling organizations to move both responsively and efficiently from concept to business outcome." This aligns with the IEEE standard definition of DevOps as "a set of principles and practices emphasizing collaboration and communication between software development teams and IT operations staff along with acquirers, suppliers, and other stakeholders in the life cycle of a software system."[12] The principles of DevOps permeate across the DevOps software pipeline, starting with the team, through release into production.

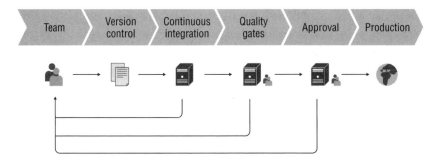

Figure B.7: DevOps Pipeline

Systems Thinking

Systems thinking is a holistic approach to viewing a system through its constituent parts as well as how those parts interrelate with one another and within the context of larger systems. The concept of systems thinking emerged in 1956, when Professor Jay Forrester founded the System Dynamics Group at MIT's Sloan School of Management. Through his experience building aircraft simulators and building computerized combat systems, he learned that the biggest impediments to problems were not on the engineering side but on the management side. He believed that this was because social systems are far more complex than physical systems. Through the System Dynamics Group, Forrester was able to mathematically model complex issues and problems.

Figure B.8: Causal Loop

Systems thinking provides us with a variety of tools and methods, such as the causal loop illustrated in Figure B.8, which allows us to model cause and effect.

Another key tool of systems thinking is the iceberg model, which begins with the observation of events or data to identify patterns over time. This approach, outlined in Figure B.9, surfaces the underlying structures that drive the events and patterns.

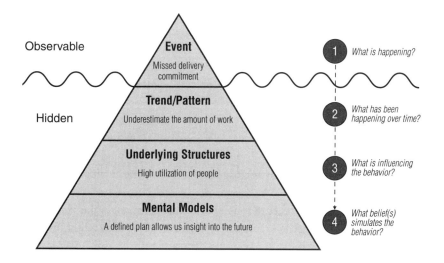

Figure B.9: Systems Thinking Iceberg

Systems thinking is a critical tool to help people solve complex problems through observation and feedback loops, allowing us to view the system as a whole instead of just the parts.

Systems Architecture

There is not a common understanding of the definition of architecture across different domains and industries, so before we tell you why architecture is important to systems, we must begin by developing a common understanding of terms such as *architecture* and *systems*.

The fifth edition of the *Shorter Oxford English Dictionary* begins by defining *architecture* as "the art or practice of designing and constructing buildings," which is not surprising, given the origins. Architecture dates back to approximately 10,000 BC, when humans moved out of caves and began building physical structures to meet a set of needs, such as shelter from the weather in areas where food was plentiful. Later, these needs became more complex. When we began growing our food, these structures needed to be built to leverage the local elements to provide safety and comfort and to store food.

Choice education further abstracted to "the complex or carefully designed structure of something."

In our case, we are designing and constructing *systems*. So, what's a system?

Going back to our trusty dictionary, a *system* is "a set of things working together as parts of a mechanism or an interconnecting network." Some examples of systems include your phone and your vehicle, but they also include things you may not have considered, such as the human brain or society as a whole. Systems can be categorized as *natural systems* or *designed systems*. For this book, we are focused on designed systems.

Given our definition of *architecture* as the complex designed structure of something and our definition of *system* as a set of things working together, we can assume that to effectively meet our needs, we need to *intentionally* design how a set of things work together. This is the basic definition of *systems architecture*.

While systems architecture may sound simple, most organizations do not invest enough in architecting their systems for speed, which includes people, processes, and tools. All systems have an architecture, but if it is not *intentional*, we will not have the culture, process, or technical architecture to meet our unique needs. The layers of the system that need to be intentionally architected include data, application, technical, culture, system, performance, and security, which are illustrated in Figure B.10.

There is agreement that organizations that deliver capabilities faster learn faster, giving them a strong competitive advantage in the market. But there is not enough discussion about how to intentionally architect our systems to enable this speed. That is what we're looking at in this appendix.

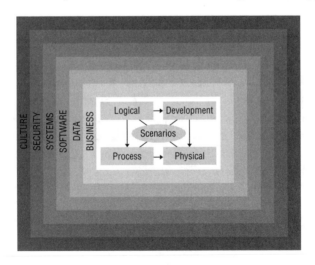

Figure B.10: Intentionally Architected Views

Systems Engineering

The International Council on Systems Engineering (INCOSE) is a not-for-profit organization founded in 1990 that works to advance the state of systems engineering (the practice of architecting designed systems). INCOSE defines engineered (designed) systems as a system designed or adapted to interact with an anticipated operational environment to achieve one or more intended purposes while complying with applicable constraints.[13]

A *basic system*, illustrated in Figure B.11, exists in an environment and has a boundary, inputs, processes, and outputs. For example, a thermometer is a basic engineered system. If a thermometer exists in Italy (*environment*) and senses the local temperature (*boundary*), that temperature is *input* through the sensor and the thermometer calculates a number (*process*) and *outputs* degrees Celsius.

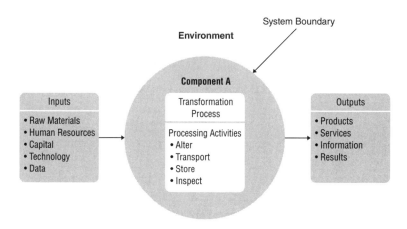

Figure B.11: Basic System

There are three basic types of systems: open, closed, and isolated. An open system can exchange both energy and matter with its environment, a closed system can exchange energy but not matter, and an isolated system is where neither energy nor matter can be exchanged. A thermometer is considered a basic *closed* system.

A *complex system* has multiple components and additional feedback loops, illustrated in Figure B.12, and requires additional work to make updates. Consider a thermometer that senses both temperature *and* humidity. It interacts with two or more components in the system and outputs readings in degrees for temperature as well as the percent of humidity in the air.

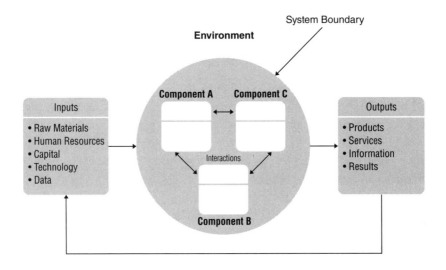

Figure B.12: Closed System with Feedback

The next level of complexity is a *system of systems*, which is a collection of systems that pool resources together to create a more complex system with increased capabilities (illustrated in Figure B.13). Let's consider the average digital thermometer that controls the heating and cooling of your home. These thermometers often have additional components that not only read the temperature but compare the temperature to low and high preset threshold settings. The thermometer not only outputs readings in degrees but also determines whether to trigger heating or air-conditioning systems based on the thresholds that have been set.

If the common thermometer that controls the heating and cooling of your house is a *complex system of systems*, how would you describe a car? A satellite? A collection of satellites? Or SpaceX's Starlink satellite internet constellation? Given the level of complexity of these systems and their criticality to our daily lives, an intentionally disciplined approach to architecture across all of the system layers is required to meet the needs of all system stakeholders and especially our customers.

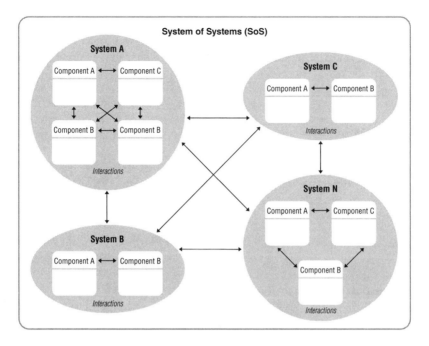

Figure B.13: System of Systems

What's an Interface?

Systems are connected through a series of system *interfaces*, which establish physical connections between systems with a common messaging syntax so that data is understood by both systems. The fundamental aspect of an interface is functional and is defined as the inputs and outputs of functions. Physical interfaces are functions that are performed by physical elements (system elements), and inputs/outputs of functions are also carried by physical elements.

Both functional and physical aspects are considered when designing interfaces. A detailed analysis of an interface shows the function "send" located in one system element, the function "receive" located in the other one, and the function "carry" as being performed by the physical interface that supports the input/output flow.

The architectural interface embraces the concept, allowing its design to be conceived as elements that run independently from each other to allow technological progress without compromising the system.

Architecture of interfaces requires extensive analysis of a variety of trade-offs including performance, scalability, reliability, availability, extensibility, maintainability, manageability, and security. There is no one best practice; it's key to understand your system of interest (SOI) and the context in which it will exist.

Conceptual Models

A systems engineer will start the architecture process with a series of conceptual models that help the engineer to reason out the types of structures and behaviors that are necessary for the system they are designing. The models include (in order) logical architecture, process architecture, development architecture, physical architecture, and scenarios and use cases:

Logical architecture: This is a representation of the structure of the system technology concepts. The goal is to be able to communicate the architecture of the system and understand the intent without having to make specific technology choices. Logical architecture can be used to describe the functionality of the system to the end user. The two most common diagrams utilized for logical architecture are class diagrams (which represents the structure of the system) and state diagrams (which represents the behavior of the system).

Process architecture: After the logical architecture has been established, systems engineers produce a process definition. Process diagrams communicate dynamic aspects of the system. Common diagrams utilized in process architecture are sequence diagrams (process flow in a time sequence), communication diagrams (interactions between objects), and activity diagrams (workflow with steps and actions).

Development architecture: The development views can be used to share how engineers plan to implement the functionality of the system. The most common diagrams for development are component diagrams (wiring of the software components) and package diagrams (demonstrates dependencies). Package diagrams are an excellent way to communicate dependencies and share the hierarchy of elements.

Physical architecture: This architecture communicates the physical aspect of the systems, such as the connections between components. Deployment diagrams are commonly used to demonstrate where capabilities are located physically and the execution of the architecture of the system.

Scenarios and use cases: Lastly, scenarios and use cases demonstrate the architecture in action. Use cases communicate a flow of activities and typically have multiple personas to demonstrate multiple paths through the system. The paths through the system can be aligned with the backlog, with each path representing one or more features.

Business Architecture

According to Red Hat, "Business architecture is a foundational practice that bridges the gaps between business and technology."[14] Business architecture originated in the 1980s in cross-organizational design but has evolved to become a first-class citizen in the enterprise architecture frameworks.

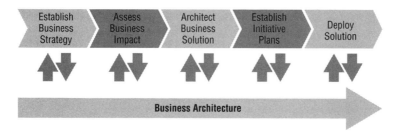

Figure B.14: Business Architecture

Business architecture provides a line of sight from business objectives to execution. This transparency enables alignment across the organization, which is important to be able to deliver value to our stakeholders. How many times have you seen different portions of your organization working toward conflicting goals? The lack of transparency creates excessive waste, hindering delivery.

Data Architecture

While there is no single definition of data architecture, for our purposes, we will define *data* as facts and statistics collected together for reference or analysis and *architecture* as the complex designed structure of something. Therefore, data architecture is the organization of facts and statistics to be utilized for reference or analysis throughout the life cycle.

Over the years, data has moved from siloed and on-premises solutions to open, cloud-based offerings based on the needs of the business, the technology available, and the constraints of the system, as illustrated in Figure B.15. Key considerations in developing the best data architecture include what your budget is, where the sources of your data are, what types of data you have (structured, semistructured, unstructured), what regulations you have on where data is located, how your data will be used (reporting, streaming analytics), and how you need to present the data.

Data architecture exists on a spectrum where one side of the spectrum for data architecture is *on-premises*, *siloed*, and *consolidated data within silos* and the other end of the spectrum is *cloud-based*, *open*, and *federated*. Given society's need for more knowledge faster, it's likely your business needs to continuously move to cloud-based, open, and federated for optimal performance and speed.

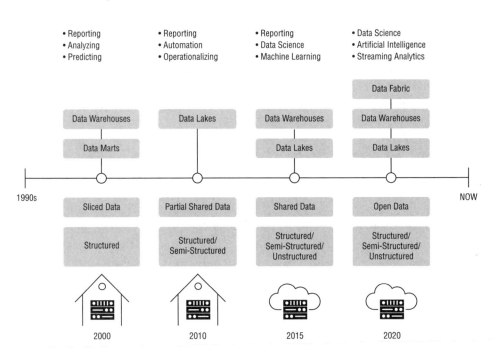

Figure B.15: Data Architecture Evolution

Software/Application Architecture

For our discussion, we are going to use TechTarget's definition of *application architecture*, which they define as a structural map of how an organization's software applications are assembled and how those applications interact with each other to meet business or user requirements.[15] The application architecture evolution illustrated in Figure B.16 has been evolving since the 1980s. Application architectures have evolved from monolithic architecture, which is a single unified unit; to service-oriented architecture, which decomposes the single unit into a set of modules; to microservices, which further decomposes into decoupled units of capability; and finally to serverless, which runs those discrete microservices on demand. Just as with the other architecture domains, the trend is toward modular and loosely coupled capability.

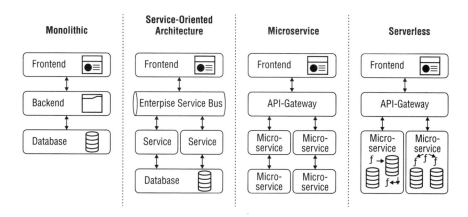

Figure B.16: Evolution of Software Architecture

Software architecture intentionally defines how things connect to ensure we have a scalable, reliable, and available solution. You need to invest in your architecture if the knowledge to build the system exists only in the *wetware* of your teams. Patterns to consider when architecting your application architecture include layered, client-server, event-driven, microkernel application, and microservices:

Layered architecture: Software components are separated into layers of work. The most common is N-tier architecture, which typically has four layers (presentation, business, application, and data). This is a great option if you are building web applications.

Client-server architecture: This pattern has a server and multiple entities. The clients make requests, and the server responds. This pattern is ideal for banking applications.

Event-driven architecture: This pattern involves services that are triggered by an event, such as user interaction. This is an excellent pattern for websites.

Microkernel application architecture: This pattern involves a core application with plug-in modules. This pattern is best for product-based or scheduling applications.

Microservices architecture: This pattern basically builds small services and aggregates them like building blocks to create larger solutions. This pattern can be used on a range of websites to real-time embedded systems.

Design Thinking

Similar to systems thinking, design-thinking concepts were formulated in the 1950s and evolved in the 1960s with what was referred to as *design science*. Design thinking is a human-centered approach to developing products and services. The second wave of design thinking came with Nigel Cross, a human-computer interaction researcher, who published "Designerly Ways of Knowing." The 1990s continued to evolve design thinking concepts with the publishing of "Wicked Problems in Design Thinking."[16]

There are six steps in the design thinking process: *empathize*, *define*, *ideate*, *prototype*, *test*, and *implement* (illustrated in Figure B.17). The steps allow designers to immerse themselves in their customers' experiences to provide better products.

We have all experienced products that were not intuitive and created problems instead of solving them. The body of knowledge built around design thinking allows everyone to be creative designers, instead of a rare few, with a few proven principles defined next:

1. **Empathize:** Understand the problem from your customer's viewpoint through observation and engagement.
2. **Define the problem:** Create a clear statement of the problem we are trying to solve.
3. **Ideate:** Brainstorm solution ideas.
4. **Prototype:** Build a tactile, low-cost solution to the problem.
5. **Test:** Get feedback from your prototyped solution.
6. **Implement:** Build the final solution.

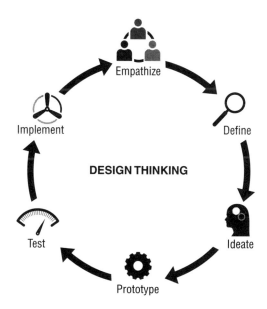

Figure B.17: Six Steps of Design Thinking

Harvard Business Review describes how people are rooted in status quo or behavioral norms that block our ability to imagine new possibilities.[17] Design thinking has allowed many businesses to overcome these challenges to build superior solutions.

Digital Engineering

According to the Software Engineering Institute, digital engineering is defined as an "integrated digital approach that uses authoritative sources of systems data and models as a continuum across disciplines to support life cycle activities from concept through disposal."[18] Digital engineering encompasses a number of methods and tools, including modeling, digital twins, augmented reality, virtual reality, and quantum computing, to name just a few. Digital engineering has exploded across various domains, from aerospace and automotive to energy and health care. The common denominator is that they have to reduce the cost and schedule of products while maintaining safety and security.

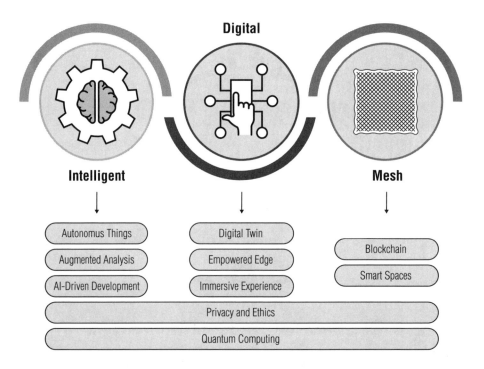

Figure B.18: Digital Engineering

The US Department of Defense (DoD) builds some of the most complex bespoke systems in the world. To that end, they have adopted a digital engineering approach to modernizing how the DoD designs, develops, delivers, operates, and sustains systems. In 2019, Philomena Zimmerman from the Office of the Under Secretary of Defense for Research and Engineering wrote extensively about the DOD's Digital Engineering Strategy and outlined its goals, outlined in Figure B.19. The goals of digital engineering are to provide increased visibility, accuracy, and adaptability for physical systems and their stakeholders.

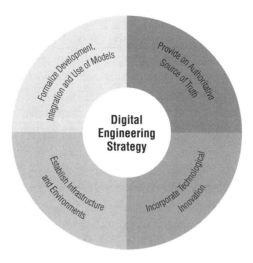

Figure B.19: DoD Digital Engineering Goals

Agile Manufacturing

Lean principles have been well grounded in manufacturing for decades. Lean focuses on reducing waste and improving flow and sustaining it. Sustaining products to keep their value requires a continuous improvement mindset, another core component of Lean. Henry Ford was also interested in how to improve flow and productivity. This led him to implement and use the conveyor belt on the assembly line, which resulted in significant gains in manufacturing production. The Lean community is entrenched in improving flow with additional practices that enable flexibility and adaptability.

The combination of Lean with Agile principles has shifted parts of manufacturing to focus not only on flow and the delivery of products but on how to build better systems that are faster, more flexible, and more adaptive. This is known as Agile manufacturing and eXtreme manufacturing, which are similar in principle.

eXtreme manufacturing started in 2006 with Joe Justice of Wikispeed. Wikispeed designed and manufactured cars for the road and the racetrack. Wikispeed set records for speed of development and won on of the most famous car races in the world, the Nürburgring 24 Hours in Germany.

The Agile practices used included modular architecture, mob development, and eXtreme programming from the software world. These practices have evolved into a set of principles and practices for hardware and in a domain area with high safety requirements and a highly regulated environment. This approach to manufacturing products creates an environment with digital tools and designs, providing the ability to adapt to new priorities and technologies quickly. Justice has since gone on to work at Tesla and shares Agile hardware practices globally with books, blogs, keynotes, conferences, classes, interviews, and more.

In the publication *Agile Manufacturing*, Nicola Accialini defines this concept as "being able to offer a greater production mix using fewer resources."[19] Agile manufacturing includes faster product development cycles and the ability to develop the factory as an Agile manufacturing system. The smart factory is a cyber-physical system. As a result, we are using cyber-physical systems (aka the factory) to build cyber-physical systems (e.g., cars, spacecraft).

The Agile manufacturing system is reconfigurable quickly "according to the production mix and market demands, with high-level autonomy."[20] The ability to achieve mass customization in the factory is built on the foundation and mindset of innovation, training, collaboration, and leadership enabled through three pillars: Modularity and Flexibility, Lean and Six Sigma, and Automation/Industry 5.0, as depicted by Accialini in Figure B.20.

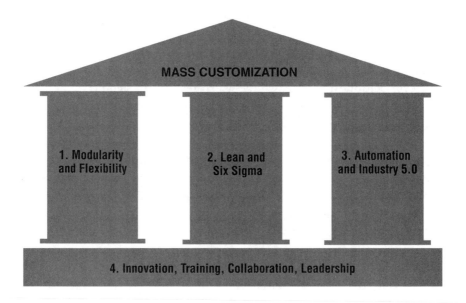

Figure B.20: The Three Pillars of Agile Manufacturing Systems
Source: Nicola Accialini, Agile Manufacturing.
Used with permission.

Additive Manufacturing

Additive manufacturing (AM) is defined by ASTM International as "a process of joining materials to make objects from 3D model data, usually layer upon layer, as opposed to subtractive manufacturing methodologies."[21] Additive manufacturing began in the late 1980s with stereolithography (SLA) from 3D Systems, a process that solidifies thin layers of ultraviolet (UV) light-sensitive liquid polymer using a laser. This process of incrementally building up a three-dimensional object in layers is also referred to as 3D printing. The 3D printing process is illustrated below in Figure B.21.

The 3D printing steps are:

1. Create a 3D model.
2. Convert the 3D model into an .STL file that can be sliced into thin layers.
3. Transfer the .STL file to the 3D printer.
4. Set up the machine (3D printer) with configuration parameters and materials.
5. Build the product per motion coding.
6. Remove the product from the printer.
7. Complete post-processing tasks (cleaning, polishing, painting).

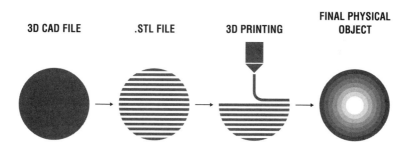

Figure B.21: 3D Printing Steps

Additive manufacturing differs from traditional manufacturing in incrementally adding materials to build a product. In contrast, traditional manufacturing removes materials to build a product. While additive manufacturing has many benefits, the speed of prototype development and the speed of prototype for production allow a much faster feedback loop for physical systems.

APPENDIX C

TOOLS: QUICK REFERENCE GUIDE

Throughout this book, many tools and techniques have been shared that can help you along your journey. While our list is not all inclusive, it provides a foundational set that has been implemented in many organizations. Table C.1 summarizes the Industrial DevOps principles, several of the relevant tools or techniques, and when to use them.

Table C.1: Quick Reference Guide

Principle	Tool or Technique	When to Use
Principle 1: Organize for the Flow of Value	Organizational structure analysis	Pick the best structure for your organizational goals.
	Value stream mapping	Define the mapping of flow from customer need to product delivery. To find bottlenecks in flow.
	Team Topologies composition	Defining the team structures to consider different team types.
Principle 2: Apply Multiple Horizons of Planning	Lean canvas	Capture business needs, a solution summary, and benefits.
	Road mapping	Define high-level goals over time and connect strategy to execution.
	Patterns of decomposition	Decompose the system functionality to fit into timeboxes.

Principle	Tool or Technique	When to Use
Principle 3: Implement Data-Driven Decisions	Objective and Key Results	Set high-level strategic objectives for the organization with defined targeted results.
	Flow Metrics	Identify, measure, and analyze the workflow through your system.
	Variety of digital tools such as 3D printing, prototypes, digital and simulated models, digital shadows, digital twins, and emulators.	Decide on the tools needed to shift testing and manufacturing left. May be part of investment planning as tools evolve.
Principle 4: Architect for Change and Speed	Modular Open Systems Approach (MOSA)	Enhance interoperability, flexibility, scalability, and affordability by promoting modularity and standardization .
	Artificial Intelligence and Machine Learning	Allow architects and systems engineers to evaluate multiple design options with multiple parameters and remove bad options quickly while amplifying engineering knowledge. Improve predictive analytics and forecasting.
	Digital Twin	Reduce risk by testing in a virtual environment before deploying new capabilities. Improve predictive maintenance.
Principle 5: Iterate, Manage Queues, Create Flow	Flow Charts and Visualization Tools	Visualize the flow of the system and find bottlenecks
	Experimentation toward a Solution	Isolate bottlenecks. GAIL iterative learning and feedback. Test alternate design decisions.
	Set-Based Design	Explore multiple sets of possibilities at the subsystem level against broad targets and proactively explore the limits of hardware design.

Principle	Tool or Technique	When to Use
	Kanban	Visualize and manage queues and the flow of work through the system.
Principles 6: Establish Cadence and Synchronization for Flow	Cadence Analysis (consider factors such as team size, availability, complexity of work, and the need for coordination and feedback)	Determine team and program cadence for planning and demonstrations.
	Program Calendar of Events	Schedule recurring planning and demonstration events for the next six months to a year out.
Principle 7: Integrate Early and Often	CI/CD Pipelines	Automate and streamline product development process.
	Incremental Integration	Detect integration issues and provides opportunities for frequent testing and feedback.
Principle 8: Shift Left	Build Labs Early and Evolve (e.g., software-in-the-loop, hardware-in-the-loop, digital twins)	Enable early and incremental development and testing with improved quality.
	Behavior-Driven Development	Ensure you are building the right thing and that you are building the thing right.
	Shift-Left Manufacturing	Enable regular feedback loops to improve verification in hardware and manufacturing design and reduce rework further downstream. Improve quality, which reduces cost of rework.
	Testing Strategy	Define your testing approach and invest in the digital environments required to meet quality standards and data needs.

Principle	Tool or Technique	When to Use
Principle 9: Apply a Growth Mindset	Cultural Surveys	Determine current cultural beliefs.
	Intentionally Architect Culture	Drive the organization to implement new behaviors
	Recognition Program	Provide a program where peers can recognize and elevate each other's successes.
	Psychological Safety	Create an environment where people feel free to speak up, share their ideas, share risks, and share failures, which results in increased innovations, learning, and successes.
	T-Shaped Skills	Help teams to deliver faster and fill gaps. T-shaped skills and cross-domain learning are critical.
	Learning Strategy	Shape a common language and shared mental models across leaders, functions, and teams as the new practices become part of the organization's culture.
	Coaching	Help the team and organization improve their practices to deliver value with measurable results while building high-performing teams. This can be Lean coaches, Agile coaches, or leaders/managers as coaches.

Bibliography

"2022 Global Space Industry Report." *Benchmark International* (blog). September 23, 2022. https://blog.benchmarkcorporate.com/2022-global -space-industry-report.

Accialini, Nicola. *Agile Manufacturing: Strategies for Adaptive, Resilient and Sustainable Manufacturing*. Self-published, December 2, 2022.

Afifi-Sabet, Keumars. "How BMW Embraced Agile to Hit New Speeds." IT Pro. August 9, 2018. https://www.itpro.com/agile-development/31552/how-bmw -embraced-agile-to-hit-new-speeds.

Agile Alliance. "Subway Map to Agile Practices." AgileAlliance.org (website). Accessed May 21, 2023. https://www.agilealliance.org/agile101/subway-map-to-agile -practices/.

Agile Manifesto. "Manifesto for Agile Software Development." AgileManifesto.org (website). Accessed May 1, 2023. https://agilemanifesto.org/.

———. "Principles behind the Agile Manifesto." AgileManifesto.org (website). Accessed May 1, 2023. https://agilemanifesto.org/principles.html.

Amazon. "Who We Are." AboutAmazon.com (website). Accessed May 7, 2023. https:// www.aboutamazon.com/about-us.

Amazon Web Services. "Formula 1 Redesigns Car for Closer Racing and More Exciting Fan Experience by Using AWS HPC Solutions." AWS.Amazon.com (website). 2021. https://aws.amazon.com/solutions/case-studies/formula-1-graviton2/.

Anderson, Katie. *Learning to Lead, Leading to Learn: Lessons from Toyota Leader Isao Yoshino on a Lifetime of Continuous Learning*. California: Integrand Press, 2020.

Anderson, Patrick. "Code on the Road: BMW's Global Value Stream Management (VSM) Journey." *Plainview Blog*. January 27, 2021. https://blog.planview.com /code-on-the-road-bmws-global-value-stream-management-vsm-journey/.

Apple, Inc. "Apple Inc. Form 10-K for the Fiscal Year Ended September 30, 2017." Accessed May 7, 2023. https://s2.q4cdn.com/470004039/files/doc_ financials/2017/10-K_2017_As-Filed.pdf.

Apposite Technologies. "Satellite Testing: Best Practices for Application Performance Testing Over Satellite." Accessed May 14, 2023. https://www.apposite-tech.com/ blog/best-practices-to-test-application-perfor mance-over-satellite-networks/.

Augustine, Sanjiv, Roland Cuellar, and Audrey Scheere. *From PMO to VMO: Managing for Value Delivery*. Oakland, CA: Berrett-Koehler Publishers, 2021.

Automotive World. "Bosch: Did You Know. . . .." *Automotive World* (website). July 21, 2020. https://www.automotiveworld.com/news-releases/bosch-did-you-know/.

Bellan, Rebecca. "Joby Aviation's Contract with US Air Force Expands to include Marines." *TechCrunch*. August 10, 2022. https://techcrunch.com/2022/08/10/joby-aviations-contract-with-u-s-air-force-expands-to-include-marines/.

Berg, Cliff. "SpaceX's Use of Agile Methods." Medium. December 9, 2019. https://cliffberg.medium.com/spacexs-use-of-agile-methods-c63042178a33.

Biggs, Renee. "What Is a Business Architect and How Do You Become One?" Red Hat. December 16, 2022. https://www.redhat.com/architect/business-architect-career#:~:text=Business%20architecture%20is%20a%20foundational,gaps%20between%20business%20and%20technology.&text=A%20business%20architect%20is%20a,the%20business%20and%20its%20users.

Blinde, Loren. "Lockheed Martin Completes First LM 400 Multi-MissionSpacecraft." Intelligence Community News. January 31, 2023. https://intelligence communitynews.com/lockheed-martins-completes-first-lm-400-multi-mission-spacecraft/.

Bloomberg Technology. "SpaceX Nails Landing of Reusable Rocket on Land." June 30, 2021. YouTube video, 1:00. https://www.youtube.com/watch?v=Aq7rDQx9jns.

Bosch. "We Are Bosch." Accessed May 7, 2023. https://www.wearebosch.com/index.en.html.

Brenton, Flint. "What Is Value Stream Management? A Primer For Enterprise Leadership." *Forbes*. July 8, 2019. https://www.forbes.com/sites/forbes techcouncil/2019/07/08/what-is-value-stream-management-a-primer-for-enterprise-leadership/?sh=1ebb073e7b67.

Brooks, Shilo. "Why Did the Wright Brothers Succeed When Others Failed?" Scientific American. March 14, 2020. https://blogs.scientificamerican.com/observations/why-did-the-wright-brothers-succeed-when-others-failed/.

Buchanan, Richard. "Wicked Problems in Design Thinking." *Design Issues* 8, no. 2 (1992): 5–21.

Carlos, Juan. "SpaceX: Enabling Space Exploration through Data and Analytics." Harvard Business School Digital Innovation and Transformation. March 23, 2021. https://d3.harvard.edu/platform-digit/submission/spacex-enabling-space-exploration-through-data-and-analytics/.

Carignan, Louis-Philippe, "Agile, Is It Just a Delivery Mechanism?" Scrum.org (web site). August 27, 2014. https://www.scrum.org/resources/blog/agile-it-just-delivery-mechanism.

Carr, David F. "Digital Transformation Leaders' Secret: Dump Traditional Org Charts." The Enterprisers Project. January 3, 2020. https://enterprisersproject.com/article/2020/1/digital-transformation-leaders-how-organize.

Chevron. "The Chevron Way: Getting Results the Right Way." Accessed May 7, 2023. https://www.chevron.com/about/the-chevron-way.

Cleland-Huang, Jane, Ankit Agrawal, Michael Vierhauser, and Christoph Mayr-Dorn. "Visualizing Change in Agile Safety-Critical Systems." *IEEE Software* 38, no. 3 (June 2020): 43–51.

Comella-Dorda, Santiago, Lavkesh Garg, Suman Thareja, and Belkis Vasquez-McCall. "Revisiting Agile Teams after an Abrupt Shift to Remote." McKinsey & Company. April 28, 2020. https://www.mckinsey.com/capabilities/people-and-organizational-performance/our-insights/revisiting-agile-teams-after-an-abrupt-shift-to-remote.

Crook, Joy, Suzette Johnson, Harry Koehnemann, Jeffrey Shupack, Steven J. Spear, Hasan Yasar, Robin Yeman, Jeff Boleng, Eileen Wrubel, and Mik Kersten. *Applied Industrial DevOps 2.0: A Hero's Journey*. IT Revolution DevOps Enterprise Forum. 2020.

Daily, Jim, and Jeff Peterson. "Predictive Maintenance: How Big Data Analysis Can Improve Maintenance." In *Supply Chain Integration Challenges in Commercial Aerospace*. December 2016. https://link.springer.com/chapter/10.1007/978-3-319-46155-7_18.

Datta, Ganesh. "Best of 2022: How DORA Metrics Can Measure and Improve Performance." DevOps.com. December 30, 2022. https://devops.com/how-dora-metrics-can-measure-and-improve-performance/.

Deming, W. Edwards. *Out of the Crisis*. Cambridge, MA: The MIT Press, 2000.

Deutschman, Alan. "Change or Die." *Fast Company*. May 5, 2005. https://www.fastcompany.com/52717/change-or-die.

Devalekar, Ashish. "Why High Employee Engagement Results in Accelerated Revenue Growth." *Forbes*. July 14, 2021. https://www.forbes.com/sites/forbesbusinesscouncil/2021/07/14/why-high-employee-engagement-results-in-accelerated-revenue-growth/?sh=6b1f5952597b.

Digital.ai. *15th Annual State of Agile Report*. Accessed May 22, 2023. https://digital.ai/resource-center/analyst-reports/15th-state-of-agile-report/.

Digital.ai. *16th Annual State of Agile Report*. Accessed April 30, 2023. https://info.digital.ai/rs/981-LQX-968/images/AR-SA-2022-16th-Annual-State-Of-Agile-Report.pdf.

Digital Engineering 24/7. "Simulation Advances NASCAR Race Car Development." November 5, 2020. https://www.digitalengineering247.com/article/simulation-advances-nascar-race-car-development/.

Dikert, Kim, Maria Paasivaara, and Casper Lassenius. "Challenges and Success Factors for Large-Scale Agile Transformations: A Systematic Literature Review." *Journal of Systems and Software* 119 (2016): 87–108.

Doerr, John. *Measure What Matters: How Google, Bono, and the Gates Foundation Rock the World with OKRs*. New York: Portfolio/Penguin, 2018.

Donahue, Colum. "How to Keep Pace with Digital Transformation and Avoid Becoming Obsolete." *Forbes*. February 12, 2020. https://www.forbes.com/sites/theyec/2020/02/12/how-to-keep-pace-with-digital-transformation-and-avoid-becoming-obsolete/?sh=1ee1f6024861.

Dweck, Carol S. *Mindset: The New Psychology of Success*. New York: Ballantine Books, 2006.

Edmondson, Amy C. "Strategies for Learning from Failure." *Harvard Business Review*. April 2011. https://hbr.org/2011/04/strategies-for-learning-from-failure.

Edwards, Damon, Ben Grinnell, Gary Gruver, Suzette Johnson, Kaimar Karu, Terri Potts, and Rosalind Radcliffe. *Making Matrixed Organizations Successful with DevOps: Tactics for Transformation in a Less Than Optimal Organization*. IT Revolution DevOps Enterprise Forum. 2017.

Errick, Kirsten. "Senate Committee Approves AGILE Procurement Act for IT and Communications Tech." Nextgov. August 4, 2022. https://www.nextgov.com

/it-modernization/2022/08/senate-committee-approves-agile-procurement-act-it-and-communications-tech/375404/.

European Space Agency. "Galileo Clock Anomalies under Investigation." January 19, 2017. https://www.esa.int/Applications/Navigation/Galileo _clock_anomalies _under_investigation.

European Space Agency. "No 33–1996: Ariane 501—Presentation of Inquiry Board Report." July 23, 1996. https://www.esa.int/Newsroom/Press_Releases/Ariane_501 _-_Presentation_of_Inquiry_Board_report.

Feldscher, Jacqueline. "China Could Overtake US in Space without 'Urgent Action,' Warns New Pentagon Report." Defense One. August 24, 2022. https://www .defenseone.com/technology/2022/08/china-could-overtake-us-space-without -urgent-action-report/376261/.

Ferguson, Kevin. "Application Architecture." TechTarget. October 2021. https://www .techtarget.com/searchapparchitecture/definition/application-architecture.

Ferreira, Soraia. "Warming to Her Task: How Soraia Ferreira is Giving Heating Technology a Digital Boost." Bosch. Accessed May 1, 2023. https://www.bosch.com /stories/warming-to-her-task/.

Field, Kyle. "Tesla Has Applied Agile Software Development to Automotive Manufacturing." CleanTechnica. September 1, 2018. https://cleantechnica .com/2018/09/01/tesla-has-applied-agile-software-development-to-automotive -manufacturing/.

"Flow Metrics—Understanding Value Stream Metrics." Flow Framework. November 8, 2022. https://flowframework.org/flow-metrics/.

Forbes Technology Council. "11 Benefits of Behavior-Driven Product Development." *Forbes*. August 9, 2019. https://www.forbes.com/sites/forbestechcouncil /2019/08/09/11-benefits-of-behavior-driven-product-development/?sh =75cd9bef1a3c.

Fowler, Martin, and Jim Highsmith. "The Agile Manifesto." *Software Development*. August 2001. http://www.hristov.com/andrey/fht-stuttgart/The_Agile_Manifesto_ SDMagazine.pdf.

Fritzsch, Jonas, Justus Bogner, Markus Haug, Anna Cristina Franco Da Silva, Carolin Rubner, Matthias Saft, Horst Sauer, and Stefan Wagner. "Adopting Microservices and DevOps in the Cyber-Physical Systems Domain: A Rapid Review and Case Study." *Software Practice and Experience* 53, no. 3 (2023): 790–810.

Ford. "The Moving Assembly Line and the Five-Dollar Workday." Accessed May 21, 2023. https://corporate.ford.com/articles/history/moving-assembly-line .html#:~:text=What%20made%20this%20assembly%20line,built%20step %2Dby%2Dste.

Forsgren, Nicole, Jez Humble, and Gene Kim. *Accelerate: The Science of Lean Software and DevOps: Building and Scaling High Performing Technology Organizations*. Portland, OR: IT Revolution, 2018.

Furuhjelm, Jörgen, Johan Segertoft, Joe Justice, and J. J. Sutherland. "Owning the Sky with Agile: Building a Jet Fighter Faster, Cheaper, Better with Scrum." Scrum Inc. 2017. https://www.scruminc.com/wp-content/uploads/2015/09/Release-version_ Owning-the-Sky-with-Agile.pdf.

Gartner. "Use Test-First Development Processes to Jump-Start Your SDLC." December 18, 2019. https://www.gartner.com/en/documents/3978408.

Gemba Academy. "Hoshin Planning." Accessed May 21, 2023. https://www.gemba academy.com/resources/gemba-glossary/hoshin-planning.

GlobalCom. "The Cost of Building and Launching a Satellite." Accessed May 21, 2023. https://globalcomsatphone.com/costs/.

Gouré, Dan. "DoD's Move to the Cloud Is Critical to Operate at the 'Speed of Relevance.'" RealClear Defense. June 25, 2018. https://www.realcleardefense.com/articles/2018/06/25/dods_move_to_the_cloud_speed_of_relevance.html.

Grant, Adam. *Think Again: The Power of Knowing What You Don't Know*. New York: Viking, 2021.

Groysberg, Boris, Jeremiah Lee, Jesse Price, and J. Yo-Jud Cheng. "The Leader's Guide to Corporate Culture." *Harvard Business Review*.January–February 2018. https://hbr.org/2018/01/the-leaders-guide-to-corporate-culture.

Gruver, Gary, and Tommy Mouser. *Leading the Transformation: Applying Agile and DevOps Principles at Scale*. Portland, OR: IT Revolution Press, 2015.

Hamdi, Shabnam, Abu Daud Silong, Zoharah Binti Omar, and Roziah Mohd Rasdi. "Impact of T-Shaped Skill and Top Management Support on Innovation Speed: The Moderating Role of Technology Uncertainty." *Cogent Business & Management* 3, no. 1 (March 2016).

Harter, Jim. "Is Quiet Quitting Real?" Gallup. September 6, 2022. https://www.gallup .com/workplace/398306/quiet-quitting-real.aspx.

Harvard Business Review Analytic Services. *Competitive Advantage through DevOps: Improving Speed, Quality, and Efficiency in the Digital World*. January 25, 2019. https://hbr.org/resources/pdfs/comm/google/CompetitiveAdvantageThrough DevOps.pdf.

Heistand, Christopher, Justin Thomas, Nigel Tzeng, Andrew R. Badger, Luis M. Rodriguez, Aaron Dalton, Jesse Pai, Austin Bodzas, and Derik Thompson. "DevOps for Spacecraft Flight Software." *2019 IEEE Aerospace Conference* (March 2019): 1–16.

Hessing, Ted. "History of Lean." Six Sigma Study Guide. Accessed May 21, 2023. https://sixsigmastudyguide.com/history-of-lean/.

Howard, Ben. "What Is Agile Aerospace? Learn Planet's Approach." Planet. October 16, 2019. https://www.planet.com/pulse/what-is-agile -aerospace-learn-planets -approach/.

Howell, Elizabeth. "8 Ways that SpaceX has Transformed Spaceflight." Space.com. March 25, 2022. https://www.space.com/ways-spacex -transformed-spaceflight.

Humble, Jez, and David Farley. *Continuous Delivery: Reliable Software Releases through Build, Test, and Deployment Automation*. Boston: Addison-Wesley, 2010.

IEEE Standards Association. "IEEE 2675-2021: IEEE Standard for DevOps: Building Reliable and Secure Systems Including Application Build, Package, and Deployment." April 2016. https://standards.ieee.org/ieee/2675/6830/.

INCOSE. "Systems and SE Definitions." INCOSE.org (website). Accessed July 13, 2023. https://www.incose.org/about-systems-engineering/system-and-se-definition/system-and-se-definitions.

International Center for Clinical Excellence. "The Backwards Bicycle: What It Takes to Unlearn Old and Learn New Habits." July 21, 2021. YouTube video, 3:56. https://www.youtube.com/watch?v=CqwfGUhYBEA.

International Organization for Standardization (ISO). "ISO 26262-1:2011 Road Vehicles—Functional Safety—Part 1: Vocabulary." November 2011. https://www.iso.org/standard/43464.html.

Ismail, Nick. "AI Solutions Required for Fast-Paced Application Development?" *Information Age*. July 5, 2018. https://www.information-age.com/ai-assistant-application-development-10672/#:~:text=Gartner%20predicts%20that%20%E2%80%9Cby%202022,what%E2%80%99s%20called%20low%2Dcode%20development.

ISO Update. "What Is Quality." September 23, 2019. https://isoupdate.com/resources/what-is-quality/.

ISS National Laboratory. "From Sample to Results: In-Space Data Analysis Enables Quicker Data Return." April 5, 2022. https://www.issnationallab.org/hpe-sbc2-inspace-data-analysis-enables-quicker-return/.

Johnson, Suzette, Robin Yeman, Harry Koehnemann, Jeffrey Shupack, Matt Aizcorbe, Adrian Cockcroft, Michael Mckay, and Hasan Yasar. *Building Industrial DevOps Stickiness: Applying Insights*. Portland, OR: IT Revolution Press, 2021.

Johnson, Suzette, Diane Lafortune, Dean Leffingwell, Harry Koehnemann, Stephen Magill, Steve Mayner, Avigail Ofer, Anders Wallgren, Robert Stroud, and Robin Yeman. *Industrial DevOps: Applying DevOps and Continuous Delivery to Significant Cyber-Physical Systems*. IT Revolution DevOps Enterprise Forum. 2018.

Johnson, Suzette, Robin Yeman, Harry Koehnemann, Jeffrey Shupack, Hasan Yasar, Ben Grinnell, Deborah Brey, Steve Farley, and Josh Corman. "Overcoming Barriers to Industrial DevOps: Working with the Hardware-Engineering Community." *The DevOps Enterprise Journal* 4, no. 2 (Fall 2022).

Jurkiewicz, Jakub, Ramanathan Yegyanarayanan, Myles Hopkins, Atulya Krishna Mishra, and Sandeep P R. *Whitepaper: Employee Engagement*. Business Agility Institute. December 5, 2019. https://businessagility.institute/learn/whitepaper-employee-engagement/275.

Justice, Joe. *Scrum Master: The Agile Training Seminar for Business Performance (Agile Business Performance from the Agile Business Institute Book 1)*. Self-published, February 4, 2021. Kindle.

Kalenda, Martin, Petr Hyna, and Bruno Rossi. "Scaling Agile in Large Organizations: Practices, Challenges, and Success Factors." *Journal of Software: Evolution and Process* 30, no. 10 (2018).

Kennedy, Michael. *Product Development for the Lean Enterprise: Why Toyota's System Is Four Times More Productive and How You Can Implement It*. Richmond, VA: The Oaklea Press, 2008.

Kersten, Mik. *Project to Product: How to Survive and Thrive in the Age of Digital Disruption with the Flow Framework*. Portland, OR: IT Revolution Press, 2018.

Khan, Faisal. "Evolution of Mars Rovers in the Past 25 Years." Medium. August 12, 2022. https://medium.com/technicity/evolution-of-mars-rovers-in-the-past-25-years-b89ec4fbc43f.

Kharpal, Arjun. "'Pathetic' Performance Has Left U.S. 'Well Behind' China in 5G Race, Ex-Google CEO Eric Schmidt Says." CNBC. February 17, 2022. https://www.cnbc.com/2022/02/17/us-well-behind-china-in-5g-race-ex-google-ceo-eric-schmidt-says.html.

Kim, Gene, Jez Humble, Patrick Debois, and John Willis. *The DevOps Handbook: How to Create World-Class Agility, Reliability, and Security in Technology Organizations.* Portland, OR: IT Revolution Press, 2016.

Klender, Joey. "Tesla Makes $9,500 per Car, Eight Times as Much as Toyota." TESLARATI. November 9, 2022. https://www.teslarati.com/tesla-toyota-profit-margin-8-times-q3-2022/#:~:text=This%20is%20where%20the%20financials,report%20states%20this%20is%20unconfirmed.

Kohn, Alfie. "Why Incentive Plans Cannot Work." *Harvard Business Review.* September–October 1993. https://hbr.org/1993/09/why-incentive-plans-cannot-work.

Kolodny, Lora. "Joby Aviation Can't Hit Production Targets on Time, According to Short Sellers' Report." CNBC. September 30, 2022. https://www.cnbc.com/2022/09/30/joby-aviation-cant-hit-production-targets-on-time-short-report.html#:~:text=Joby%27s%20projections,by%20the%20end%20of%202024.

Kontoghiorghes, Constantine, Susan M. Awbre, and Pamela L. Feurig. "Examining the Relationship between Learning Organization Characteristics and Change Adaptation, Innovation, and Organization Performance." *Human Resource Development Quarterly* 16, no. 2. (June 22, 2005). 185–212. https://onlinelibrary.wiley.com/doi/abs/10.1002/hrdq.1133.

Kotter, John. "Leading Change: Why Transformation Efforts Fail." *Harvard Business Review.* May–June 1995. https://hbr.org/1995/05/leading -change-why -transformation-efforts-fail-2.

Kowzan, Mateusz, and Patrycja Pietrzak. "Continuous Integration in Validation of Modern, Complex, Embedded Systems." *2019 IEEE/ACM International Conference on Software and System Processes* (2019): 160–164.

Kumar, S. Annand, and R. V. S. Prasad. "Chapter 2 - Basic Principles of Additive Manufacturing," in Additive Manufacturing, edited by M. Manjaiah, K. Raghavendra, N. Balashanmugam, and J. Paulo Davim. Woodhead Publishing, 2021.

LaBerge, Laura, Clayton O'Toole, Jeremy Schneider, and Kate Smaje. "How COVID-19 Has Pushed Companies over the Technology Tipping Point—And Transformed Business Forever." McKinsey & Company. October 5, 2020. https://www.mckinsey.com/capabilities/strategy-and-corporate-finance/our-insights/how-covid-19-has-pushed-companies-over-the-technology-tipping-point-and-transformed-business-forever.

Landau, Peter. "What Is Lead Time? How to Calculate Lead Time in Different Industries." ProjectManager.com (website). March 21, 2023. https://www.projectmanager.com/blog/lead-time-how-to-calculate.

Lambert, Fred. "First Look at Tesla's New Structural Battery Pack that Will Power Its Future Electric Cars." Electrek. January 19, 2021. https://electrek.co/2021/01/19/tesla-structural-battery-pack-first-picture/.

Lawton, C. A. "The History of Lean – Part 1." C. A. Lawton Co. July 10, 2019. https://calawton.com/lean-history-part-1/.

Lean Enterprise Institute. "Cycle Time." Accessed May 25, 2023. https://www.lean.org/
lexicon-terms/cycle-time/.

Lean Enterprise Institute. "Value Streams." Accessed May 5, 2023. https://www.lean
.org/lexicon-terms/value-stream/.

Lean Enterprise Institute. "What Is Lean?" Accessed May 21, 2023. https://www.lean
.org/explore-lean/what-is-lean/.

Liedtka, Jeanne. "Why Design Thinking Works." *Harvard Business Review*. September–
October 2018. https://hbr.org/2018/09/why-design -thinking-works.

Lofgren, Eric. "How Saab Uses Agile Principles to Develop the Gripen Fighter."
Acquisition Talk (blog). November 17, 2021. https://acquisitiontalk.com/2021/11/
how-saab-uses-agile-principles-to-develop-the-gripen-fighter/.

Mahmoud, Magdi S., and Yuanqing Xia. "Chapter 13—Secure Estimation Subject
to Cyber Stochastic Attacks." In *Cloud Control Systems: Analysis, Design and
Estimation*, edited by Stephen Ison and Lucy Budd, 373–404. Cambridge, MA:
Academic Press, 2020. https://www.sciencedirect.com/science/article/abs/pii/
B9780128187012000214.

Marquet, L. David. *Turn the Ship Around! A True Story of Turning Followers into Leaders*.
New York: Penguin, 2012.

Maxwell, John C. *Leadershift: The 11 Essential Changes Every Leader Must Embrace*.
Nashville: HarperCollins Leadership, 2019.

McDermott, Rose. "Prospect Theory." Britannica. Accessed May 21, 2023. https://
www.britannica.com/topic/prospect-theory/additional-info#history.

Mersino, Anthony. "Why Agile Is Better than Waterfall (Based on Standish Group
Chaos Report 2020)." Vitality Chicago. November 1, 2021. https://vitalitychicago.
com/blog/agile-projects-are-more-successful-traditional-projects/.

Mochari, Ilan. "Why Half of the S&P 500 Companies Will Be Replaced in the Next
Decade." *Inc*. March 23, 2016. https://www.inc.com/ubs/five-lessons-from-
companies-that-are-making-a-difference.html.

Moorhead, Patrick. "The Past and Present Are Telling of the Future, So What's
Next for Planet?" *Forbes*. November 29, 2021. https://www.forbes.com/sites/
patrickmoorhead/2021/11/29/the-past-and-present-are-telling-of-the-future-so-
whats-next-for-planet/?sh=35f3376e41e4.

Morris, David. *Storm on the Horizon: Khafji—The Battle That Changed the Course of the
Gulf War*. New York: Presidio Press, 2005.

MotorBiscuit. "3 Toyotas with the Lowest 10-Year Maintenance Costs." December 3,
2022. https://www.motorbiscuit.com/3-toyotas-lowest-10-year-maintenance-costs/.

Muruganandham, Shivaprakash. "Cloud Computing: Ratcheting the Satellite
Industry Forward." NSR. June 15, 2020. https://www.nsr.com/cloud-computing
-ratcheting-the-satellite-industry-forward/.

NASA Science Mars Exploration. "NASA Readies Perseverance Mars Rover's Earthly
Twin." September 4, 2020. https://mars.nasa.gov/news/8749/nasa-readies
-perseverance-mars-rovers-earthly-twin/.

Nath, Shyam Varan, Ann Dunkin, Mahesh Chowdhary, and Nital Patel. *Industrial
Digital Transformation: Accelerate Digital Transformation with Business Optimization,
AI, and Industry 4.0*. Birmingham, UK: Packt Publishing, 2020.

National Science Foundation. "Cyber-Physical Systems (CPS)." January 7, 2021. https://new.nsf.gov/funding/opportunities/cyber-physical-systems-cps.

National Science Foundation. "Cyber-Physical Systems: Enabling a Smart and Connected World." Accessed May 1, 2023. https://www.nsf.gov/news/special_reports/cyber-physical/.

National Science Foundation. "Cyber-Physical Systems (CPS): PROGRAM SOLICITATION NSF 21-551." Accessed May 21, 2023. https://www.nsf.gov/pubs/2021/nsf21551/nsf21551.htm.

Nelson, Mark. "Beyond the Buzzword: What Does Data-Driven Decision-Making Really Mean?" *Forbes*. September 23, 2022. https://www.forbes.com/sites/tableau/2022/09/23/beyond-the-buzzword-what-does-data-driven-decision-making-really-mean/?sh=7f7d2e2725d6.

Nolan, Dennis P., and Eric T. Anderson. *Applied Operational Excellence for the Oil, Gas, and Process Industries*. Waltham, MA: Gulf Professional Publishing, 2015.

Office of Nuclear Energy. "Idaho National Laboratory Demonstrates First Digital Twin of a Simulated Microreactor." July 14, 2022. https://www.energy.gov/ne/articles/idaho-national-laboratory-demonstrates-first-digital-twin-simulated-microreactor.

Office of the Under Secretary of Defense for Acquisition and Sustainment. *Agile Software Acquisition Guidebook-Best Practices & Lessons Learned from the FY18 NDAA Section 873/874 Agile Pilot Program*. Washington, D.C.: Feb. 27, 2020. See National Defense Authorization Act for Fiscal Year 2018, Pub. L. No. 115-91, §§ 873-874 (2017) (codified at 10 U.S.C. §§ 2223a note, 2302 note)· Section 874 of the NDAA for Fiscal Year 2018.

O'Shaughnessy, Robert. "NASA's Software Development Gets Complex." Federal News Network. December 14, 2022. https://federalnewsnetwork.com/technology-main/2022/12/nasas-software-development-gets-complex/.

Palucka, Tim. "NASCAR's Next Gen Race Car Proven Safe by Simulation." InfoWorld. October 19, 2022. https://www.infoworld.com/article/3676588/nascar-s-next-gen-race-car-proven-safe-by-simulation.html.

Perez, Carlotta. *Technological Revolutions and Financial Capital: The Dynamics of Bubbles and Golden Ages*. Cheltenham, UK: Edward Elgar, 2003.

Phipps, Blaine E., Michael H. Young, and Nathan G. Christensen. "Structural Optimization Helps Launch Space Payloads." *Concept to Reality*. Altair.com. Summer/Fall 2011. https://altair.com/docs/default-source/resource-library/c2r2011-structural-opt.pdf?sfvrsn=33f8a9fc_3.

Pink, Daniel. *Drive: The Surprising Truth About What Motivates Us*. Edinburgh, UK: Canongate Books, 2010.

Pollock, Roy V. H., Andy Jefferson, and Calhoun W. Wick. *The Six Disciplines of Breakthrough Learning: How to Turn Training and Development into Business Results*. Hoboken, NJ: John Wiley & Sons, 2015.

Porsche AG. "Agile Transformation: Bringing the Porsche Experience into the Digital Future with SAFe." Medium.com (wesbite). May 11, 2021. https://medium.com/next-level-german-engineering/agile-transformation-bringing-the-porsche-experience-into-the-digital-future-with-safe-bf3df9bdbd08

———. "A Journey through the history of Porsche: Why Agile Work Isn't New to Us." Medium.com (website). January 16, 2020. https://medium.com/next-level-german-engineering/a-journey-through-the-history-of-porsche-why-agile-work-isnt-new-to-us-9664ab30f035

Prior, Madeleine. "Launcher Successfully Tests Its 3D Printed E2 Rocket Engine in Full Thrust." 3Dnatives. April 30, 2022. https://www.3dnatives.com/en/launcher-3d-printed-e2-rocket-engine-in-full-thrust-300420224/#!.

Rajadurai, Niroshan. "How SpaceX Develops Software." Coders Kitchen. June 25, 2020. https://www.coderskitchen.com/spacex-software-development-and-testing/.

Ready, Douglas A., Carol Cohen, David Kiron, and Benjamin Pring. "The New Leadership Playbook for the Digital Age." *MIT Sloan Management Review*. January 2020. https://www.cognizant.com/en_us/insights/documents/the-new-leadership-playbook-for-the-digital-age-codex5350.pdf.

Reichmann, Kelsey. "In a First for the DoD, Kubernetes Installed on U-2 Dragon Lady." Avionics International. October 9, 2020. https://www.aviationtoday.com/2020/10/09/first-dod-kubernetes-installed-u-2-dragon-lady/.

Reinertsen, Donald G. *The Principles of Product Development Flow: Second Generation Lean Product Development*. Redondo Beach, CA: Celeritas Pub, 2009.

Reiterer, Stefan H., Sinan Balci, Desheng Fu, Martin Benedikt, Andreas Soppa, and Helena Szczerbicka. "Continuous Integration for Vehicle Simulations." *2020 25th IEEE International Conference on Emerging Technologies and Factory Automation* (2020): 1023–1026.

Relativity Space. "World's Largest 3D Metal Printers: Rockets Built for the Future." Retrieved May 12, 2023. https://www.relativityspace.com/stargate.

Riedel, Robin. "Rideshares in the Sky by 2024: Joby Aviation Bets Big on Air Taxis." McKinsey & Company. November 19, 2021. https://www.mckinsey.com/industries/aerospace-and-defense/our-insights/rideshares-in-the-sky-by-2024-joby-aviation-bets-big-on-air-taxis.

Ries, Eric. *The Lean Startup: How Today's Entrepreneurs Use Continuous Innovation to Create Radically Successful Businesses*. New York: Crown Business, 2011.

———. *The Startup Way: How Modern Companies Use Entrepreneurial Management to Transform Culture and Drive Long-Term Growth*. New York: Currency, 2017.

Rigby, Darrell, Jeff Sutherland, and Hirotaka Takeuchi. "Embracing Agile: How to Master the Process That's Transforming Management." *Harvard Business Review*. May 2016. https://hbr.org/2016/05/embracing-agile.

Rigby, Darrell, Jeff Sutherland, and Hirotaka Takeuchi. "The Secret History of Agile Innovation." *Harvard Business Review*. April 20, 2016. https://hbr.org/2016/04/the-secret-history-of-agile-innovation.

Risher, Howard. "Daniel Pink's 'Zen of Compensation' Is Anything But." Compensation Café. June 8, 2012. https://www.compensationcafe.com/2012/06/daniel-pinks-zen-of-compensation-is-anything-but.html.

Rocket Lab. "Responsive Space: Accelerating the Path to Orbit with Rapid Call-Up Launch on Demand and Agile Satellite Solutions." Accessed May 21, 2023. https://www.rocketlabusa.com/launch/responsive-space/.

Root, Al. "How Much Is SpaceX Worth Now? More Than Lockheed Martin."
Barron's. December 13, 2022. https://www.barrons.com/articles/
spacex-lockheed-martin-market-value-51670942386.

Rosenberg, Barry. "DevStar for B-21 Design, Build, Sustainment." Breaking Defense.
June 30, 2021. https://breakingdefense.com/2021/06/devstar-for-b-21-design
-build-sustainment/.

Rother, Mike. *Toyota Kata: Managing People for Improvement, Adaptiveness, and Superior
Results*. New York: McGraw-Hill Education, 2010.

Royce, Winston W. "Managing the Development of Large Software Systems." *ICSE '87:
Proceedings of the 9th International Conference on Software Engineering* (March 1981):
328–338.

Sacolick, Isaac. "3 Ways DevOps Can Support Continuous Architecture." InfoWorld.
May 30, 2022. https://www.infoworld.com/article/3662290/3-ways-devops-can-
support-continuous-architecture.html.

Santos, Paula de Oliveira, and Marly Monteiro de Carvalho. "Exploring the Challenges
and Benefits for Scaling Agile Project Management to Large Projects: A Review."
Requirements Engineering 27, no. 1 (March 2022): 117–134.

Saran, Cliff. "How Containerisation Helps VW Develop Car Software." Computer
Weekly (website). June 23, 2021. https://www.computerweekly.com/news
/252502600/How-containerisation-helps-VW-develop-car-software.

Scaled Agile Framework. "Create a Lean-Agile Center of Excellence." December 9, 2022.
https://scaledagileframework.com/lace/.

Schein, Edgar H. *Organizational Culture and Leadership*. Hoboken, NJ: John Wiley &
Sons, 2017.

Shepard, David, and Julia Scherb. "What Is Digital Engineering and How Is It Related
to DevSecOps?" Software Engineering Institute. November 16, 2020. https://
insights.sei.cmu.edu/blog/what-digital-engineering-and-how-it-related-devsecops/.

Shieber, Jonathan. "Wondering About Getting a Job at SpaceX? Elon Musk Says
Innovation Is the Main Criterion." TechCrunch. February 28, 2020. https://
techcrunch.com/2020/02/28/wondering-about-getting-a-job-at-spacex-elon-musk-
says-innovation-is-the-main-criterion/.

Shook, John. "How to Change a Culture: Lessons from NUMMI." *MIT Sloan
Management Review*. January 1, 2010. https://sloanreview.mit.edu/article/
how-to-change-a-culture-lessons-from-nummi/.

Simon, Herbert A. *Administrative Behavior: A Study of Decision-Making Processes in
Administrative Organization*. New York: The Free Press, 1997.

Simon, Herbert A., George B. Dantzig, Robin Hogarth, Charles R. Plott, Howard Raiffa,
Thomas C. Schelling, Kenneth A. Shepsle, Richard Thaler, Amos Tversky, and Sidney
Winter. "Decision Making and Problem Solving." *Interfaces* 17, no. 5 (October 1987):
11–31.

Sinek, Simon. *The Infinite Game*. New York: Portfolio/Penguin, 2019.

Singh, Victor, and K. E. Willcox. "Engineering Design with Digital Thread." *American
Institute of Aeronautics and Astronautics (AIAA) Journal* 56, no. 11 (November 2018).

Skelton, Matthew, and Manuel Pais. *Team Topologies: Organizing Business and
Technology Teams for Fast Flow*. Portland, OR: IT Revolution, 2019.

Smart, Jonathan. *Sooner Safer Happier: Antipatterns and Patterns for Business Agility.* Portland, OR: IT Revolution Press, 2020.

Somers, Richard J., James A. Douthwaite, David J. Wagg, Neil Walkinshaw, and Robert M. Hierons. "Digital-Twin-Based Testing for Cyber–Physical Systems: A Systematic Literature Review." *Information and Software Technology* 156 (2023).

SpaceX. "Mars & Beyond: The Road to Making Humanity Multiplanetary." Accessed May 7, 2023. https://www.spacex.com/human-spaceflight/mars/.

———. "The Future of Design." September 5, 2013. YouTube video, 3:48. https://www.youtube.com/watch?v=xNqs_S-zEBY.

Spiegel, Rob. "Simulation Brings Design Speed to NASCAR." Design News. January 31, 2021. https://www.designnews.com/design-software/simulation-brings-design-speed-nascar.

Sprovieri, John. "BMW Applies AI to Assembly." ASSEMBLY. June 29, 2022. https://www.assemblymag.com/articles/97139-bmw-applies-ai-to-assembly#:~:text=%E2%80%9CThe%20BMW%20Group%20has%20been,production%20at%20%5Bour%5D%20plants.%20%3C/Citation%3E.

Straits Research. "CubeSat Market: Information by Size (0.25U to 1U, 1U to 3U), Application (Space Observation), Subsystem (Payloads, Structures) End-Users, and Region—Forecast till 2030." Accessed May 22, 2023. https://straitsresearch.com/report/cubesat-market.

Stray, Viktoria, Bakhtawar Memon and Lucas Paruch. "Systemic Literature Review on Agile Coaching and the Role of the Agile Coach." International Conference on Product-Focused Software Process Improvement. October 2020. https://www.researchgate.net/publication/345062509_A_Systematic_Literature_Review_on_Agile_Coaching_and_the_Role_of_the_Agile_Coach.

Strickland, Ashley. "'Rail Cars' of Material Released after NASA Spacecraft Hit Asteroid." CNN. December 15, 2022. https://www.cnn.com/2022/12/15/world/dart-mission-momentum-results-scn/index.html.

Stumpf, Rob. "At $631B, Tesla Is Now Worth More than the Next Top 6 Car Companies Combined." The Drive. December 30, 2020. https://www.thedrive.com/news/38485/at-631b-tesla-is-now-worth-more-than-the-next-top-6-car-companies-combined

Sutherland, J. J., and Joe Justice. "Agile In Military Hardware: How the SAAB Gripen became the world's most cost effective military aircraft." Scrum Inc. 2017.

Sutherland, Jeff, and J. J. Sutherland. *Scrum: The Art of Doing Twice the Work in Half the Time.* New York: Crown Business, 2014.

Szondy, David. "Lockheed Martin Develops Smart Satellite That Can Be Reprogrammed In Orbit." New Atlas. March 22, 2019. https://newatlas.com/lockheed-martin-smart-satellite/58957/.

Tao, Fei, Meng Zhang, and A. Y. C. Nee. *Digital Twin Driven Smart Manufacturing.* London: Academic Press, 2019.

Takeuchi, Hirotaka, and Ikujiro Nonaka. "The New New Product Development Game." *Harvard Business Review.* January 1986. https://hbr.org/1986/01/the-new-new-product-development-game.

Tracy, Brian. "Goals Mastery for Personal and Financial Success." BrianTracy.com. Accessed May 14, 2023. https://www.briantracy.com/files/pages/goals/mastery/lp-long.html?cmpid=2269&amp%3Bproid=2031.

Ullman, David G., and Joshua Tarbutton. "Scrum for Hardware and Systems Development." Machine Design. May 29, 2019. https://www.machinedesign.com/3d-printing-cad/article/21837829/scrum-for-hardware-and-systems-development.

US Government Accountability Office. *Agile Assessment Guide: Best Practices for Agile Adoption and Implementation.* September 2020. https://www.gao.gov/assets/gao-20-590g.pdf.

———. *F-35 Joint Strike Fighter: Assessment Needed to Address Affordability Challenges.* April 2015. https://www.gao.gov/assets/gao-15-364.pdf.

———. *Software Development: Effective Practices and Federal Challenges in Applying Agile Methods.* July 27, 2012. https://www.gao.gov/products/gao-12-681.

———. *Weapons Systems Annual Assessment: Challenges to Fielding Capabilities Faster Persist.* June 2022. https://www.gao.gov/assets/gao-22-105230.pdf.

Vacanti, Daniel S. *Actionable Agile Metrics for Predictability: An Introduction.* Self-published, March 4, 2015. Leanpub.

Verma, Pranshu. "How the 3D-Printing Community Worldwide Is Aiding Ukraine," *The Washington Post.* June 12, 2022. https://www.washingtonpost.com/technology/2022/06/12/3d-printers-ukraine-war-supplies/.

Viguié, Arnaud, and Joe Justice. "Tesla Agile Success." Agile Business Institute. Accessed May 1, 2023. https://en.abi-agile.com/tesla-agile-success/.

Wall, Mike. "SpaceX Launches 1st Test Satellites for Starlink Internet Constellation along with Spain's Paz." Space.com (website). February 22, 2018. https://www.space.com/39755-spacex-used-rocket-launches-internet-satellites.html.

Wang, Brian. "Tesla's Fully Agile Rapid Innovation." Next Big Future. December 27, 2021. https://www.nextbigfuture.com/2021/12/174206.html.

Westrum, Ron. "A Typology of Organisational Cultures." *Qual Saf Health Care* 13, no. 2 (December 2004): 22–27.

Wikipedia. "Queueing Theory." Wikipedia.com (website). Last modified April 19, 2023. https://en.wikipedia.org/wiki/Queueing_theory.

Williams, Taffy. *Think Agile: How Smart Entrepreneurs Adapt in Order to Succeed.* New York: AMACOM, 2015.

Windley, Phil. "Creating an Agile Culture through Trust and Ownership: An Interview with Pollyanna Pixton and Niel Nickolaisen." InformIT. April 10, 2014. https://www.informit.com/articles/article.aspx?p=2191027.

Witthuhn, Bri Flynn. "The Neuroscience behind Habit Change." *Forbes.* February 11, 2020. https://www.forbes.com/sites/ellevate/2020/02/11/the-neuroscience-behind-habit-change/?sh=742691976f6a.

WORKERBASE. "Benefits of Agile Manufacturing for Smart Factories." Accessed May 1, 2023. https://workerbase.com/post/benefits-of-agile-production-systems-for-smart-factories#:~:text=For%20the%20company%2C%20an%20agile,by%20improving%20productivity%20and%20OEE.

Notes

Preface

1. Johnson et al., *Industrial DevOps*.

Introduction

1. Digital.ai, *16th Annual State of Agile Report*.
2. Harvard Business Review Analytic Services, *Competitive Advantage through DevOps*.
3. Mersino, "Why Agile Is Better than Waterfall."
4. Mersino, "Why Agile Is Better than Waterfall."
5. Accialini, *Agile Manufacturing*, 26.
6. Gouré, "DoD's Move to the Cloud Is Critical to Operate at the 'Speed of Relevance.'"
7. Gouré, "DoD's Move to the Cloud Is Critical to Operate at the 'Speed of Relevance.'"
8. Benchmark International, "2022 Global Space Industry Report."
9. Harvard Business Review Analytic Services, *Competitive Advantage through DevOps*.
10. Perez, *Technological Revolutions and Financial Capital*.
11. US Government Accountability Office, *Weapons Systems Annual Assessment*.
12. Kharpal, "'Pathetic' Performance Has Left U.S. 'Well Behind' China in 5G Race, Ex-Google CEO Eric Schmidt Says."
13. Feldscher, "China Could Overtake US in Space without 'Urgent Action,' Warns New Pentagon Report."
14. Ries, *The Lean Startup*.
15. Donahue, "How to Keep Pace with Digital Transformation and Avoid Becoming Obsolete."
16. Harter, "Is Quiet Quitting Real?"
17. Mersino, "Why Agile Is Better than Waterfall."
18. Mersino, "Why Agile Is Better than Waterfall."
19. Porsche AG, "A Journey through the History of Porsche."
20. Porsche AG, "Agile Transformation."
21. Afifi-Sabet, "How BMW Embraced Agile to Hit New Speeds."
22. Afifi-Sabet, "How BMW Embraced Agile to Hit New Speeds."
23. Kersten, *Project to Product*, 190.
24. Field, "Tesla Has Applied Agile Software Development to Automotive Manufacturing."
25. Viguié and Justice, "Tesla Agile Success."
26. Wang, "Tesla's Fully Agile Rapid Innovation."
27. Wang, "Tesla's Fully Agile Rapid Innovation."
28. Wang, "Tesla's Fully Agile Rapid Innovation."
29. Johnson et al., *Industrial DevOps*.
30. Deming, *Out of the Crisis*.

Chapter 1

1. Nath et al., *Industrial Digital Transformation*.
2. Ross, as quoted in: Carr, "Digital Transformation Leaders' Secret."
3. LaBerge et al., "How COVID-19 Has Pushed Companies over the Technology Tipping Point—and Transformed Business Forever."
4. "Cyber-Physical Systems (CPS)," National Science Foundation.
5. "Cyber-Physical Systems," National Science Foundation.
6. Mahmoud and Xia, "Chapter 13—Secure Estimation Subject to Cyber Stochastic Attacks," in *Cloud Control Systems*, edited by Ison and Budd, 373–404.
7. Musk, as quoted in: Shieber, "Wondering About Getting a Job at SpaceX?"
8. Viguié and Justice, "Tesla Agile Success."
9. Viguié and Justice, "Tesla Agile Success."
10. Stumpf, "At $631B, Tesla Is Now Worth More than the Next Top 6 Car Companies Combined."
11. Berg, "SpaceX's Use of Agile Methods."
12. Root, "How Much Is SpaceX Worth Now? "
13. Howard, "What Is Agile Aerospace?"
14. Ferreira, "Warming to Her Task."
15. Furuhjelm et al., "Owning the Sky with Agile."
16. O'Shaughnessy, "NASA's Software Development Gets Complex."
17. US Government Accountability Office, *Weapons Systems Annual Assessment*.

Chapter 2

1. Mochari, "Why Half of the S&P 500 Companies Will Be Replaced in the Next Decade."
2. Harvard Business Review Analytic Services, *Competitive Advantage*.
3. Wang, "Rapid Innovation."
4. Gruver and Mouser, *Leading the Transformation*, 15.
5. de Oliveira Santos and Monteiro de Carvalho, "Exploring the Challenges and Benefits for Scaling Agile Project Management to Large Projects," 117–134.
6. Klender "Tesla Makes $9,500 per Car, Eight Times as Much as Toyota."
7. Smart et al., *Sooner Safer Happier*.
8. Harvard Business Review Analytic Services, *Competitive Advantage*.
9. Sutherland and Sutherland, *Scrum*.
10. Rigby, Sutherland, and Takeuchi, "Embracing Agile."
11. Rigby, Sutherland, and Takeuchi, "Embracing Agile."
12. Jurkiewicz et al., *Whitepaper: Employee Engagement*.
13. Devalekar, "Why High Employee Engagement Results in Accelerated Revenue Growth."
14. Rigby, Sutherland, and Takeuchi, "Embracing Agile."
15. "Benefits of Agile Manufacturing for Smart Factories," WORKERBASE.

Chapter 3

1. Johnson et al., "Overcoming Barriers to Industrial DevOps."
2. "Principles behind the Agile Manifesto," The Agile Manifesto.
3. "Manifesto for Agile Software Development," The Agile Manifesto.
4. Marquet, *Turn the Ship Around*, 161.

5. "Principles behind the Agile Manifesto," The Agile Manifesto.
6. Comella-Dorda et al., "Revisiting Agile Teams after an Abrupt Shift to Remote."
7. Cleland-Huang et al., "Visualizing Change in Agile Safety-Critical Systems," 43–51.
8. Johnson et al., "Overcoming Barriers to Industrial DevOps."

Chapter 4

1. Edwards et al., *Making Matrixed Organizations Successful with DevOps*.
2. Skelton and Pais, *Team Topologies*.
3. "Value Streams," Lean Enterprise Institute.
4. Justice, personal communication with the authors, 2022.
5. Windley, "Creating an Agile Culture through Trust and Ownership."
6. Ullman and Tarbutton, "Scrum for Hardware and Systems Development"; Furuhjelm et al., "Owning the Sky"; Sutherland and Justice, "Agile in Military Hardware."
7. Furuhjelm et al., "Owning the Sky."
8. Anderson, "Code on the Road."
9. Anderson, "Code on the Road."
10. Kersten, *Project to Product*, 83.
11. Kersten, *Project to Product*, 78.
12. Anderson, "Code on the Road."

Chapter 5

1. Royce, "Managing the Development of Large Software Systems," 328–338.
2. Royce, "Managing the Development of Large Software Systems," 328–338.
3. Morris, *Storm on the Horizon*.
4. Reinertsen, *The Principles of Product Development Flow*.
5. "Mars & Beyond," SpaceX.
6. "We Are Bosch," Bosch.
7. "The Chevron Way," Chevron.
8. "Apple Inc. Form 10-K for the Fiscal Year Ended September 30, 2017," Apple, Inc.
9. "Our Missions and Values," NASA.
10. "Who We Are," Amazon
11. Ries, *The Startup Way*.
12. Wall, "SpaceX Launches 1st Test Satellites for Starlink Internet Constellation along with Spain's Paz."
13. Lofgren, "How Saab Uses Agile Principles to Develop the Gripen Fighter."
14. Lofgren, "How Saab Uses Agile Principles to Develop the Gripen Fighter."
15. US Government Accountability Office, *F-35 Joint Strike Fighter*; Lofgren, "How Saab Uses Agile Principles to Develop the Gripen Fighter"; Furuhjelm et al., "Owning the Sky."
16. Lofgren, "How Saab Uses Agile Principles to Develop the Gripen Fighter."

Chapter 6

1. Nelson, "Beyond the Buzzword."
2. Reinertsen, *Principles of Product Development Flow*.
3. Lofgren, "How Saab Uses Agile Principles to Develop the Gripen Fighter."
4. Sutherland and Justice, "Agile in Military Hardware."
5. Singh and Willcox, "Engineering Design with Digital Thread."
6. Cleland-Huang et al., "Visualizing Change."

7. Somers et al., "Digital-Twin-Based Testing for Cyber-Physical Systems."
8. Joe Justice, personal communication with the authors, 2022.
9. Augustine, Cuellar, and Scheere, *From PMO to VMO.*
10. Doerr, *Measure What Matters.*
11. Kersten, *Project to Product*, 111.
12. "What Is Quality," ISO Update.
13. "Flow metrics," Flow Framework.
14. Kersten, *Project to Product.*
15. Lean Enterprise Institute, "Cycle Time."
16. Landau, "What Is Lead Time?
17. Lean Enterprise Institute, "Cycle Time."
18. Vacanti, *Actionable Agile Metrics for Predictability.*
19. Kersten, *Project to Product*, 138.
20. Datta, "Best of 2022: How DORA Metrics Can Measure and Improve Performance."
21. Forsgren, Humble, and Kim, *Accelerate*, 17.
22. Somers et al., "Digital-Twin-Based Testing."
23. Spiegel, "Simulation Brings Design Speed to NASCAR"; Palucka, "NASCAR's Next Gen Race Car Proven Safe by Simulation"; "Simulation Advances NASCAR Race Car Development," Digital Engineering 24/7.

Chapter 7

1. Szondy, "Lockheed Martin Develops Smart Satellite That Can Be Reprogrammed in Orbit."
2. Reichmann, "In a First for the DoD, Kubernetes Installed on U-2 Dragon Lady."
3. "Formula 1 Redesigns Car for Closer Racing and More Exciting Fan Experience by Using AWS HPC Solutions," Amazon Web Services.
4. Howell, "8 Ways That SpaceX Has Transformed Spaceflight."
5. Lambert, "First Look at Tesla's New Structural Battery Pack That Will Power Its Future Electric Cars."
6. Verma, "How the 3D-Printing Community Worldwide Is Aiding Ukraine."
7. "World's Largest 3D Metal Printers," Relativity Space.
8. Blinde, "Lockheed Martin Completes First LM 400 Multi-Mission Spacecraft."
9. Saran, "How Containerisation Helps VW Develop Car Software."
10. Muruganandham, "Cloud cCmputing: Ratcheting the Satellite Industry Forward."
11. "3 Toyotas with the Lowest 10-Year Maintenance Costs," MotorBiscuit.
12. Daily and Peterson, "Predictive Maintenance: How Big Data Analysis Can Improve Maintenance."
13. Fritzsch et al., "Adopting Microservices and DevOps in the Cyber-Physical Systems Domain," 790–810.
14. Sacolick, "3 Ways DevOps Can Support Continuous Architecture."
15. Sacolick, "3 Ways DevOps Can Support Continuous Architecture."
16. Rosenberg, "DevStar for B-21 Design, Build, Sustainment."
17. Sprovieri, "BMW Applies AI to Assembly."
18. "From Sample To Results: In-Space Data Analysis Enables Quicker Data Return," ISS National Laboratory.
19. Khan, "Evolution of Mars Rovers in the Past 25 Years."
20. Howard, "What Is Agile Aerospace?"
21. Moorhead, "The Past and Present Are Telling of the Future, So What's Next for Planet?"

22. Riedel, "Rideshares in the Sky by 2024."
23. Bellan, "Joby Aviation's Contract with US Air Force Expands to Include Marines."
24. Kolodny, "Joby Aviation Can't Hit Production Targets on Time, According to Short Sellers' Report."

Chapter 8

1. Brooks, "Why Did the Wright Brothers Succeed When Others Failed?"
2. Brooks, "Why Did the Wright Brothers SucceedWhen Others Failed?"
3. Harvard Business Review Analytic Services, *Competitive Advantage*.
4. Simon, *Administrative Behavior*.
5. Simon et al., "Decision Making and Problem Solving," 11–31.
6. Kennedy, *Product Development for the Lean Enterprise*, 22.
7. Kennedy, *Product Development for the Lean Enterprise*.
8. Reinertsen, *Principles of Product Development Flow*.
9. Wikipedia, "Queueing Theory."
10. Johnson et al., "Overcoming Barriers to Industrial DevOps."
11. Bloomberg Technology, "SpaceX Nails Landing of Reusable Rocket on Land."
12. Carlos, "SpaceX."
13. Carlos, "SpaceX."
14. Prior, "Launcher Successfully Tests Its 3D Printed E2 Rocket Engine in Full Thrust."
15. Phipps, Young, and Christensen, "Structural Optimization Helps Launch Space Payloads."

Chapter 10

1. "World's Largest 3D Metal Printers," Relativity Space.
2. Kim et al., *The DevOps Handbook*, 153–154.
3. Humble and Farley, *Continuous Delivery*, 58–59.
4. Kowzan and Pietrzak, "Continuous Integration in Validation of Modern, Complex, Embedded Systems," 160–164.
5. Kowzan and Pietrzak, "Continuous Integration in Validation of Modern, Complex, Embedded Systems."
6. Tao, Zhang, and Nee, *Digital Twin Driven Smart Manufacturing*.
7. Kowzan and Pietrzak, "Continuous Integration in Validation of Modern, Complex, Embedded Systems."
8. Reiterer et al., "Continuous Integration for Vehicle Simulations," 1023–1026.
9. Automotive World, "Bosch: Did You Know. . . ."

Chapter 11

1. European Space Agency, "No 33–1996: Ariane 501."
2. Ismail, "AI Solutions Required for Fast-Paced Application Development?"
3. Rajadurai, "How SpaceX Develops Software."
4. Rajadurai, "How SpaceX Develops Software."
5. Europeans Space Agency, "Galileo Clock Anomalies under Investigation."
6. Heistand et al., "DevOps for Spacecraft Flight Software," 1–16.
7. Strickland, "'Rail Cars' of Material Released after NASA Spacecraft Hit Asteroid."
8. International Organization for Standardization, "ISO 26262-1:2011 Road Vehicles — Functional Safety — Part 1."

9. Simon, "Massive Autonomous Vehicle Sensor Data: What Does It Mean?," .
10. Gartner, "Use Test-First Development Processes to Jump-Start Your SDLC."
11. Forbes Technology Council, "11 Benefits of Behavior-Driven Product Development."
12. "NASA Readies Perseverance Mars Rover's Earthly Twin," NASA Science Mars Exploration.
13. "Satellite Testing: Best Practices for Application Performance Testing Over Satellite," Apposite Technologies.
14. "Idaho National Laboratory Demonstrates First Digital Twin of a Simulated Microreactor," Office of Nuclear Energy.

Chapter 12

1. Dweck, *Mindset*.
2. Westrum, "A Typology of Organisational Cultures."
3. Risher, "Daniel Pink's 'Zen of Compensation' Is Anything But."
4. Tracy, "Goals Mastery for Personal and Financial Success."
5. Ries, *Lean Startup*.
6. Schein, *Organizational Culture and Leadership*.
7. Grant, *Think Again*.
8. Schein, *Organizational Culture*, 334.
9. Kohn, "Why Incentive Plans Cannot Work."
10. Pink, *Drive*.
11. Kotter, "Leading Change."
12. Kohn, "Why Incentive Plans."
13. Johnson et al., *Building Industrial DevOps Stickiness*.
14. Edmondson, "Strategies for Learning from Failure."
15. Anderson, *Learning to Lead, Leading to Learn*, 348.
16. Edmondson, "Strategies for Learning from Failure."
17. Pollock, Jefferson, and Wick, *The Six Disciplines of Breakthrough Learning*.
18. Pollock, Jefferson, and Wick, *The Six Disciplines of Breakthrough Learning*.
19. Deutschman, "Change or Die."
20. US Government Accountability Office, *Agile Assessment Guide*.
21. Stray, Memon, and Paruch, "Systemic Literature Review on Agile Coaching and the Role of the Agile Coach."
22. Bodell, "Why T-Shaped Teams Are the Future of Work."
23. Hamdi et al., "Impact of T-Shaped Skill and Top Management Support on Innovation Speed."
24. Schein, *Organizational Culture*.
25. Pollock, Jefferson, and Wick, *The Six Disciplines of Breakthrough Learning*.
26. Groysberg et al., "The Leader's Guide to Corporate Culture."
27. Dweck, *Mindset*, 127.
28. Groysberg et al., "Corporate Culture."
29. Crook et al., *Applied Industrial DevOps 2.0*.
30. Shook, "How to Change a Culture."
31. Shook, "How to Change a Culture."

Chapter 13

1. Augustine, Cuellar, and Scheere, *From PMO to VMO*, 37.

2. Nolan and Anderson, *Applied Operational Excellence for the Oil, Gas, and Process Industries*.
3. Crook et al., *Applied Industrial DevOps 2.0*.
4. Rother, *Toyota Kata*.
5. Ready et al., "The New Leadership Playbook for the Digital Age."
6. "Hoshin Planning," Gemba Academy.
7. Doerr, *Measure What Matters*.
8. Skelton and Pais, *Team Topologies*.
9. "Responsive Space: Accelerating the Path to Orbit with Rapid Call-Up Launch on Demand and Agile Satellite Solutions," Rocket Lab.

Chapter 14

1. International Center for Clinical Excellence, "The Backwards Bicycle."
2. Witthuhn, "The Neuroscience behind Habit Change."
3. Digital.ai, *15th Annual State of Agile Report*.
4. US Government Accountability Office, *Weapons Systems Annual Assessment*.
5. McDermott, "Prospect Theory."
6. Maxwell, *Leadershift*, 6.
7. Dikert, Paasivaara, and Lassenius, "Challenges and Success Factors for Large-Scale Agile Transformations."
8. Kalenda, Hyna, and Rossi, "Scaling Agile in Large Organizations."
9. Deming, *Out of the Crisis*.
10. Williams, *Think Agile*, 6.
11. Grant, *Think Again*, 255.
12. Digital.ai, *15th Annual State of Agile Report*.
13. US Government Accountability Office, *Weapons Systems Annual Assessment*.
14. US Government Accountability Office, *Weapons Systems Annual Assessment*.
15. Digital.ai, *15th Annual State of Agile Report*.
16. Dikert, Paasivaara, and Lassenius. "Challenges and Success Factors for Large-Scale Agile Transformations."
17. US Government Accountability Office, *Software Development*.
18. Dikert, Paasivaara, and Lassenius. "Challenges and Success Factors for Large-Scale Agile Transformations."
19. Kotter, "Leading Change."
20. "Create a Lean-Agile Center of Excellence," Scaled Agile Framework.
21. Kontoghiorghes, Awbre, and Feurig, "Examining the Relationship between Learning Organization Characteristics and Change Adaptation, Innovation, and Organization Performance."
22. Digital.ai, *15th Annual State of Agile Report*.
23. Errick, "Senate Committee Approves AGILE Procurement Act for IT and Communications Tech."
24. "Cyber-Physical Systems (CPS): PROGRAM SOLICITATION NSF 21-551," National Science Foundation.
25. Office of the Under Secretary of Defense for Acquisition and Sustainment, *Agile Software Acquisition Guidebook*.

Conclusion

1. Sinek, *The Infinite Game*.

2. Smart et al., *Sooner Safer Happier*.

Appendix A
1. "The Cost of Building and Launching a Satellite," GlobalCom.
2. "CubeSat Market," Straits Research.

Appendix B
1. "What is Lean?" Lean Enterprise Institute.
2. Lawton, "The History of Lean – Part 1."
3. Ford, "The Moving Assembly Line."
4. Ford, "The Moving Assembly Line."
5. Hessing, "History of Lean."
6. Ries, *The Lean Startup*.
7. Brenton, "What Is Value Stream Management?"
8. Rigby, Sutherland, and Takeuchi, "The Secret History of Agile Innovation."
9. Takeuchi and Nonaka, "The New New Product Development Game."
10. Fowler and Highsmith, "The Agile Manifesto."
11. Agile Alliance, "Subway Map to Agile Practices."
12. IEEE Standards Association, "IEEE 2675-2021."
13. INCOSE, "Systems and SE Definitions."
14. Biggs, "What Is a Business Architect and How Do You Become One?"
15. Ferguson, "Application Architecture."
16. Buchanan, "Wicked Problems in Design Thinking."
17. Liedtka, "Why Design Thinking Works."
18. Shepard and Scherb, "What Is Digital Engineering and How Is It Related to DevSecOps?"
19. Accialini, *Agile Manufacturing*, IX.
20. Accialini, *Agile Manufacturing*, XIX.
21. Kumar and Prasad, "Chapter 2 - Basic Principles of Additive Manufacturing."

Index

objective evidence, 82. *See also* data-driven decision making
    CubeSat development, 83
    example backlog with, 84, 85
    feedback cycles, 85
    iterative development, 86
    system-level demonstrations, 82–83
objectives and key results (OKRs), 89–90, 210
observability, 107
observe-orient-decide-act loop (OODA loop), 60. *See also* multiple horizons of planning
OKRs. *See* objectives and key results
OODA loop. *See* observe-orient-decide-act loop
organizational structure, 39, 212. *See also* Industrial DevOps framework; organizing for value
    divisional, 41
    functional, 40
    implementing data-driven decisions, 216
    inversing Conway's law, 212–213
    matrixed, 40, 41
    optimization, 212–213
    schedule, 213–216
    shifting, 42
organizing for value delivery, 13, 39, 50–53, 53–54
    BMW, 52–53
    coaching tips, 54
    flow of value, 45–50
    from functional silos to Agile value streams, 44
    manufacturing floor, 51
    organizational structure, 39–42
    Saab Aeronautics, 51–52
    streamlining value delivery, 39
    team composition, 42–44
    value stream, 45
OV-1 of Johns Hopkins Test Environment, 173, 174

pair programming, 75
PBAC. *See* policy-based access control
PC104 standard, 123
PDCA cycle. *See* plan-do-check-act cycle
PDSA cycles. *See* Plan-Do-Study-Act cycles
people-centric culture, 243
performance incentives and business outcomes, 193–195
physical architecture, 269. *See also* systems engineering
plan-do-check-act cycle (PDCA cycle), 132
Plan-Do-Study-Act cycles (PDSA cycles), 257
Planet Labs, 8–9
    Agile aerospace, 125–126
    CubeSats, 68
planning. *See also* Agile planning; multiple horizons of planning
    adaptive, 13
    in Agile product development, 28
    applying multiple horizons of, 58
    empirical, 58
    predictive to empirical, 55
    rolling wave, 69
platform-as-a-service, 114
PLE. *See* Product Line Engineering
PMI. *See* Project Management Institute

policy-based access control (PBAC), 111
predictive process control, 56
Predix, 116
process architecture, 268. *See also* systems engineering
process control models, 56
    empirical, 56
    predictive, 56
    predictive vs. empirical, 56–58
product delivery schedule, 213. *See also* Industrial DevOps framework
    accelerating, 216
    business rhythm to update with empirical data, 215
    decoupling scope from time, 214–215
    early testing and test automation, 218
    length of time remaining in, 215
    level of detail in, 215
    modernizing legacy systems with strangler pattern, 216–217
    optimizing system flow and feedback loops, 218
    resource loading approach, 215
    schedule grid, 214
    strangler pattern in action, 217
    streamlining integration, 217–218
    structure, 213–214
    team alignment, 215–216
production cycle, 255
Product Line Engineering (PLE), 113
product vision, 60, 65–66. *See also* multiple horizons of planning
Project Management Institute (PMI), 10
Prosci ADKAR model, 235
prospect theory, 230, 231

quarterly plan, 69. *See also* multiple horizons of planning
    CubeSat space ground communication use case, 70
    CubeSat team's quarterly road map, 70, 71
    goal of, 70
    quarterly road mapping, 61
queuing theory, 134–135
"quiet quitting", xxvi

Raytheon Technologies (RTX), 8
RBAC. *See* role-based access control
regulated environments, 238–239
Relativity Space, 112, 154
return on investment (ROI), 220
risk-based testing, 179
Robotic Process Automation (RPA), 4
ROI. *See* return on investment
role-based access control (RBAC), 111
rolling-wave planning, 69
Royce, Dr. Winston, 57
RPA. *See* Robotic Process Automation
RTX. *See* Raytheon Technologies

Saab Aeronautics, 9, 74
    built-in escalation path, 96
    communicating impediments at Saab, 74–76
    daily leadership cadence, 75

Acknowledgments

We have been on this journey for many years. The first time we presented together was on a panel at an Agile conference in Washington, DC, about ten years ago. The stories and experiences (and even some of the data) we shared were so similar we decided to join forces. Our thinking was that by working together and working with others across industry, it could drive us all toward better ways of working and delivering value faster to our customers. We have worked with many amazing people from across industry who have, in a variety of different ways, contributed to the knowledge and experiences captured in this book.

Thank you to our customers and teams: Many of these experiences are a result of the customers we have served and the teams we have worked with in support of those missions. We appreciate all we learned from those experiences, including the importance of collaboration, service, commitment, and delivering value to meet mission needs.

Thank you to Gene Kim: With great appreciation, we would like to thank Gene Kim. He provided us with a platform on many occasions to share our ideas. He and those in the DevOps Enterprise community have listened and provided input as we have shaped the principles of Industrial DevOps since 2018. This book is a direct result of his support.

Thank you to Leah Brown: We also want to acknowledge and offer special thanks to Leah Brown, Managing Editor at IT Revolution. She had the challenge of keeping us focused and on track! She met with us nearly every week as we wrote this book. Her time and expertise have been invaluable.

Thank you to Anna Noak: Thank you for getting us started! We appreciate the many discussions we had during the DevOps Enterprise Summit and other events as we formulated ideas for sharing our learning and experiences with others.

Thank you to the early contributors: We also want to recognize the many contributors who have specifically helped shape the principles of Industrial DevOps over the years. Several articles on Industrial DevOps were written prior to this book. We appreciate the time, energy, and expertise of these individuals. Co-authors and reviewers of the Industrial DevOps papers are: Josh Atwell, Matt Aizcorbe, Dr. Jeff Boleng, Deborah Brey, Adrian Cockcroft, Josh Corman, Joy Crook, Steve Farley, Ben Grinnell, George Haley, Steve Holt, Diane LaFortune, Dean Leffingwell, Vincent Lussenburg, Dr. Mik Kersten, Harry Koehnemann, Dr. Stephen Magill, Dr. Steve Mayner, Michael McKay, Avigail Ofer, Dantar Oosterwal, Jeffrey Shupack, Dr. Steve Spear, Robert Stroud, Cat Swetel, Anders Wallgren, Hasan Yasar, Simon Wardley, Dr. Ron Westrum, and Eileen Wrubel.

Thank you to our book review team: Special thank you to our book reviewers! Our reviewers come from a variety of backgrounds, experiences, and industries. Their insights, recommendations, and encouragement have been greatly valued. Their honest and open feedback helped us to shape the content and provide clarity where it was needed. Thank you so much for staying on this journey with us. The book reviewers include Nicola Accialini, Sanjiv Augustine, Dawn Beyer, Jeff Boleng, Nathan G. Christensen, Gabriela (Gabby) Cole, Braxton Cook, Nathaniel Crews, Carol Erikson, Jennifer Fawcett, Sandra J. Forney (DEng), Kyle Fox, Marshall Guillory, Laura Hart, Rick Hefner (PhD), Luke Hohmann, Kelli Houston, Judy Johnson, James (Jim) Judd, Joe Justice, Harry Koehnemann, Gordon Kranz, Eugene Lai, Tom McDermott, Edward Moshinsky, Andrew Olguin, Larri A. Rosser, Isaac Sacolick, Robert Scheurer, Steve Simske, Michael P. Trombley, Valerie Underwood, John Wetsch, Christopher D. Williams, and Hasan Yasar.

Thank you to our families: And we, of course, want to say thank you to our families, who have supported us along the way. Thank you so much for the encouragement and patience as we worked diligently month after month, year after year on this journey. This journey is so much better because of you.

About the Authors

Dr. Suzette Johnson

My initial experience with Agile-related practices began in the 1990s with product development for an innovative and cutting-edge technology startup company during the dot-com era. While we knew nothing about Agile frameworks, we did understand the importance of delivering value to the customer through short development cycles.

For the majority of my career I have worked for Northrop Grumman Corporation, a global aerospace, defense, and security company. While working for Northrop Grumman, my experiences with Scrum and eXtreme programming officially started in 2005 when working on a large data-centric program. I have been actively promoting these principles ever since.

As the Lean Agile transformation lead, I launched the Northrop Grumman Lean Agile Community of Practice and the Lean Agile Center of Excellence, providing resources to a workforce of over 95,000 people. Over the years, I have had the privilege to support over one hundred enterprise, federal, and Department of Defense initiatives in their adoption of Lean Agile and DevOps for improved business agility.

In my current role as a Northrop Grumman Fellow and Technical Fellow Emeritus, I am focused on the adoption of Industrial DevOps principles within the space sector.

I am an active participant in the National Defense Industrial Association (NDIA) Systems Engineering Division, NDIA ADAPT (Agile Delivery for Agencies, Programs, and Teams) working group, and the International Council on Systems Engineering (INCOSE). I also serve as a volunteer in K-12 education to share the fun and excitement of Science, Technology, Engineering, and Math (STEM) to inspire and grow our future leaders.

I received a Doctorate of Management at the University of Maryland with a dissertation focused on investigating the impact of leadership styles on software project outcomes in traditional and Agile engineering environments. I currently reside in Maryland with my family and our two lively Jack Russells. I look forward to continuing this journey as we advance and evolve in the digital age and build better systems faster.

Robin Yeman

I spent twenty-six years working at Lockheed Martin in various roles leading up to senior technical fellow building large systems including everything from submarines to satellites. I led the Agile community of practice supporting a workforce of 120,000 people. My initial experience with Lean practices began in the late '90s. In 2002, I had the opportunity to lead my first Agile program with multiple Scrum teams. After I had a couple months of experience, I was hooked and never turned back. I both led and supported Agile transformations for intelligence, federal, and Department of Defense organizations over the next two decades, and each one was more exciting and challenging than the last. In 2012, I had the opportunity to extend our Agile practices into DevOps, which added extensive automation and tightened our feedback loops, providing even larger results.

I have consulted for a range of Fortune 500 companies in highly regulated environments, enabling them to achieve the same results we experienced at Lockheed Martin. I engage in everything from automotive, pharmaceuticals, and energy to reimagining legacy to modern solutions using all of the tools in my toolbox, including Agile, DevOps, Lean, digital engineering, systems theory, design thinking, and more.

Currently, I am the Space Domain Lead at the Software Engineering Institute at Carnegie Mellon University.

I am and always will be a continuous learner. My education includes a bachelor's degree from Syracuse University in Computer Information Systems and a master's degree from Rensselaer Polytechnic Institute in Software Engineering, and I'm currently pursuing a PhD in Systems Engineering at Colorado State University, where I am working on my contribution to demonstrate empirical data of the benefits of implementing Agile and DevOps for safety-critical cyber-physical systems.